T0135603

Bibliografische Information der Deutschen Nationalbibliothek

Die Deutsche Nationalbibliothek verzeichnet diese Publikation in der
Deutschen Nationalbibliografie; detaillierte bibliografische Daten sind
im Internet über http://dnb.d-nb.de abrufbar.

ISBN 978-3-8325-2724-2

Logos Verlag Berlin GmbH
Comeniushof, Gubener Str. 47,
10243 Berlin
Tel.: +49 (0)30 42 85 10 90
Fax: +49 (0)30 42 85 10 92
INTERNET: http://www.logos-verlag.de

Two-sided projective resolutions, periodicity and local algebras

Inaugural-Dissertation

zur

Erlangung des Doktorgrades

der Mathematisch-Naturwissenschaftlichen Fakultät

der Universität zu Köln

vorgelegt von

Stefanie Küpper
aus Wuppertal

Logos Verlag Berlin

Köln 2010

Berichterstatter: Prof. Dr. Steffen König
Prof. Dr. Peter Littelmann

Tag der mündlichen Prüfung: 21. Oktober 2010

Introduction

The projective resolution of an algebra considered as a bimodule over itself is useful to compute Hochschild cohomology. There is no general version of a minimal two-sided projective resolution of an algebra. The standard resolution (see [5]), which is very classical, is far away from being minimal. Anick and Green [2] gave an explicit resolution, unfortunately only of the simple A-modules. It is not minimal in general. Bardzell [4] took this resolution and extended it, in the case A is monomial, to a resolution of A. It is even more illuminating as the original version. Thus in the monomial case we have a minimal resolution of A over its enveloping algebra.

The main results of this thesis are - in the monomial case - a complete understanding of the resolution. In the general case we have the new technique of a local algebra, which gives us information about the global resolution in a local context.

Considering any projective resolution of a module, we get four different types: finite, (eventually) periodic, almost periodic and infinite, but non-periodic. In the monomial case almost periodic resolutions are periodic. We describe the remaining three types in the monomial case in terms of Bardzell's construction. The terms *covered by S* and *connected by S* are introduced in Definition 2.1.

Theorem 2.9. Let $A = KQ/I$ be a finite dimensional monomial algebra and let I be generated by a minimal set of paths S.

(1) A has an infinite, but non-periodic resolution if and only if there are at least two cycles in the quiver of A that are covered and connected by S.

(2) A has an infinite eventually periodic resolution if and only if
   - there is at least one cycle in Q that is covered by S and
   - no pair of cycles is connected by S.

(3) A has a finite resolution if and only if there are no cycles in Q that are covered by S.

In the special case of a periodic algebra we get the following:

Theorem 2.19. Let A be a finite dimensional monomial algebra.

(1) Then A is periodic if and only if A is a product of self-injective Nakayama algebras, i.e., $A = \bigoplus_{i=1}^{t} KQ_i/KQ_i^{+l_i}$ with $Q_i = \tilde{A}_{n_i}$ and $l_i \geq 2$.

(2) The length of the period of every subalgebra $A_i = KQ_i/KQ_i^{+l_i}$ is $m_i = \min\{2\frac{n_i k}{l_i} \mid \frac{n_i k}{l_i} \in \mathbb{N}\}$.

(3) The length of the period of A is $m = \mathrm{lcm}\{m_1, \ldots, m_t\}$.

(4) Every indecomposable A-module that is not projective is periodic with a period of length m_i for some $i \in \{1, \ldots, t\}$.

I

Additionally we show that the type of the resolution is unchanged when passing from A to a factor algebra A/J where J is an homological ideal (Theorem 2.14).

In chapter 3 we introduce a new method to derive informations about the resolution. We associate a local algebra to an algebra A by gluing the vertices of the quiver of A:

Definition. Let $A = KQ/I$ be a finite dimensional basic K-algebra with $|Q_0| = n \geq 2$. Then the *local algebra* $A_{loc} = KQ_{loc}/I_{loc}$ is defined by the pullback in the diagram $(A/\mathrm{rad}A \simeq \bigoplus_{i=1}^{n} K)$:

$$
\begin{array}{ccc}
A_{loc} & \dashrightarrow & A \\
\Big\downarrow & & \Big\downarrow{\scriptstyle \pi} \\
K & \xrightarrow{(1,\ldots,1)} & A/\mathrm{rad}A
\end{array}
$$

The result is an algebra A_{loc} with one simple module. Although the quiver of A_{loc} looks rather different, the set of paths of positive length is the same as for A. Thus the difference must lie in the relations, on which we pay particular attention.

The resolution of any A_{loc}-module is completely determined by the resolution of the corresponding A-module together with the resolutions of the simple A-modules (Theorem 3.10). As a special case we have an explicit formula for the resolution of the simple A_{loc}-module S_{loc} (Theorem 3.14) and thus for the resolution of A_{loc}:

Corollary 3.15. Let (P_i, d_i) be the minimal projective resolution of the simple A_{loc}-module S_{loc} with $P_i = P_{loc}^{x_i}$. Then the following is a minimal projective resolution of A_{loc} over A_{loc}^e:

$$
\ldots (P_{loc}^e)^{x_i} \xrightarrow{d_i \otimes 1} (P_{loc}^e)^{x_{i-1}} \to \ldots \to (P_{loc}^e)^{x_1} \xrightarrow{d_1 \otimes 1} P_{loc}^e \xrightarrow{d_0 \otimes 1} A_{loc} \to 0
$$

In particular, the Anick-Green resolution is minimal for A if and only if it is so for A_{loc} (Theorem 4.11).

If two algebras have isomorphic local algebras, we call them locally equivalent. We give a description how to construct every algebra in a local equivalence class. This works by splitting one vertex into two if there is a so called splittable relation (Lemmata 3.18 and 4.1). With that we can determine whether there is a locally equivalent algebra B that has a finite or a periodic resolution. We call the resolution of A locally finite or locally periodic and so on. In the monomial case this leads to conditions for the resolution of A:

Theorem 3.23. Let $A = KQ/I$ be a finite dimensional monomial algebra.

(1) The resolution of A is locally finite if and only if each infinite associated sequence of paths contains at least one splittable relation.

II

(2) The resolution of A is locally periodic if and only if the number of infinite associated sequences of paths which contain no splittable relations is finite.

(3) The resolution of A is locally infinite if and only if the number of infinite associated sequences of paths which contain no splittable relations is infinite.

In the general case (chapter 4) we compare A with the associated monomial algebra A_{mon} introduced by [2]. For A_{mon} the resolution of Anick and Green is minimal. Moreover it looks very similar to the Anick/Green-resolution of A:

Lemma 4.5. Let $A = KQ/I$ be non-monomial and let A_{mon} be the associated monomial algebra. Let (P^i_{mon}, d^i_{mon}) denote the Anick/Green-resolution of a simple A_{mon}-module S_{mon} with $P^i_{mon} = \bigoplus_{j \in Q_0} P^{n^i_j}_{mon}(j)$, $n^i_j \geq 0$.
Then the Anick/Green-resolution of the simple A-module S that belongs to the same vertex as S_{mon} consists of the projective modules $P^i = \bigoplus_{j \in Q_0} P^{n^i_j}(j)$ and of the associated morphisms d^i.

Finally we compare the resolutions of A and of its associated monomial algebra A_{mon} and get the following inequalities:

Theorem 4.15. Let A be finite dimensional. The following hold:

(1) If the resolution of A_{mon} is locally quasi finite, then the resolution of A is locally finite.

(2) If the resolution of A is locally almost periodic, then the resolution of A_{mon} is either locally quasi periodic or locally quasi infinite.

(3) If the resolution of A_{mon} is locally quasi periodic, then the resolution of A is either locally finite or locally almost periodic.

(4) If the resolution of A is locally infinite, then the resolution of A_{mon} is locally quasi infinite.

If the resolution of Anick and Green is minimal for A, we get:

(5) If the resolution of A is locally finite, then the resolution of A_{mon} is locally finite.

(6) The resolution of A is locally almost periodic if and only if the resolution of A_{mon} is locally quasi periodic.

(7) If the resolution of A_{mon} is locally infinite, then the resolution of A is locally infinite.

Acknowledgments. The supervision by Steffen König was extraordinary. I am deeply grateful for the support, the many time he took for me and the helpful suggestions. Furthermore I want to thank the DFG for the financial support.

Contents

Chapter 1

Resolutions

The concepts of projective resolutions of modules over an algebra and of resolutions of the algebra itself are closely connected. In this chapter I would like to introduce two ways to construct resolutions. The first one is due to Bardzell [4] and is only defined for monomial algebras. After the technical definition, the construction is easy to work with, see the examples. The second one was published by David Anick and Edward Green in 1987 [2] and yields resolutions of the simple modules over an associative and finite dimensional K-algebra which does not need to be monomial. Throughout the chapter let K be a field and let $A = KQ/I$ be a quotient of the path algebra KQ for a finite quiver Q with an admissible ideal I. Thus A is an associative finite dimensional K-algebra.

1.1 Preliminaries

Let Q be a finite quiver, let $Q_0 = \{1, \ldots, n\}$ be the vertices and $Q_1 = \{\alpha_1, \ldots, \alpha_k\}$ the arrows in Q. The starting point of an arrow α_i is denoted by $s(\alpha_i)$ and its ending point is $e(\alpha_i)$. Let KQ be the path algebra viewed as a vector space over K with basis B, which is the set of all directed paths in Q. For any vertex i in Q there is a path e_i of length 0 in B such that $\sum_{i \in Q_0} e_i = 1_{KQ}$.

The multiplication in KQ is the concatenation of paths if the starting and ending points match. Since we work with left modules, the product of two paths v and w in B with $s(v) = e(w)$ is vw. Let I be an admissible ideal in KQ. This means $KQ^{+m} \subseteq I \subseteq KQ^{+2}$ for some $m \geq 2$. Here KQ^{+m} is the K vector space with basis B^{+m}, the set of all paths that have length $\geq m$. The algebra $A = KQ/I$ is finite dimensional and associative. If I is generated by a finite number of paths, say by a set $S = \{x_1, \ldots, x_n\}$, then $I = \langle S \rangle$ is called *monomial* and A is called a *monomial algebra*.

If there is an element of S that is a linear combination of paths $x = \sum k_i w_i$ with $k_i \in K$ and $w_i \in B$, we suppose x to be *uniform*. This means that there are primitive idempotents e, e' such that $exe' = x$, thus every path w_i has the same

1

starting and ending point. Otherwise we decompose x in $x = x_1 + \ldots + x_m$ with x_i uniform and get $x_1, \ldots, x_m \in S$ instead of $x \in S$. The elements of S are called the *relations* of A.

Let A-mod denote the category of finite dimensional A-modules. It is equivalent to the category $\text{rep}_K(Q, I)$ of K-linear finite dimensional *representations* of the quiver Q with respect to the ideal I (compare Lemma 2.17 or see [3]).

The indecomposable projective A-modules are $P(i) = Ae_i$ for all $i \in Q_0$. The simple modules are $S(i) = P(i)/\text{rad} P(i)$. For every module $M \in A$-mod there exists a projective module P_M and an epimorphism $\phi : P_M \to M$ with a minimal kernel; this projective module is called the *projective cover* of M. It is unique up to isomorphism.

Projective resolutions and cohomology

Definition 1.1. A *projective resolution* of an A-module M is an exact complex of projective A-modules together with an epimorphism $P_0 \xrightarrow{d_0} M$:

$$\ldots \to P_n \xrightarrow{d_n} P_{n-1} \to \ldots \to P_2 \xrightarrow{d_2} P_1 \xrightarrow{d_1} P_0 \xrightarrow{d_0} M \to 0.$$

A projective resolution is called *minimal* if $d_j : P_j \to \text{Im}\, d_j$ is a projective cover for all j as well as $d_0 : P_0 \to M$ is a projective cover.

Denote by $P(M)$ the part of the resolution without M: $\ldots \to P_n \xrightarrow{d_n} P_{n-1} \to \ldots \to P_2 \xrightarrow{d_2} P_1 \xrightarrow{d_1} P_0 \to 0$. Applying the functor $\text{Hom}_A(-, N)$ for any A-module N to the complex $P(M)$, we get

$$0 \to \text{Hom}_A(P_0, N) \xrightarrow{\text{Hom}_A(d_1, N)} \text{Hom}_A(P_1, N) \xrightarrow{\text{Hom}_A(d_2, N)} \text{Hom}_A(P_2, N) \ldots$$

The cohomology of this complex is defined to be

$$\text{Ext}_A^n(M, N) = \ker \text{Hom}_A(d_{i+1}, N)/\text{Im}\, \text{Hom}_A(d_i, N)$$

where $\text{Ext}_A^0(M, N) = \text{Hom}_A(M, N)$.

The enveloping algebra

Definition 1.2. The *opposite algebra* of A is the K-algebra A^{op} with the same underlying set and the same K-vector space structure as A. But the multiplication in A^{op} denoted by $*$ is reversed by the formula $a * b' = (ba)'$. The algebra $A^e = A \otimes_K A^{op}$ is called the *enveloping algebra* of A.

Every element of A^e is of the form $a \otimes b'$ with $a, b \in A$. The multiplication is defined by $(a \otimes b')(c \otimes d') = ac \otimes (b' * d') = ac \otimes (db)'$. So the unit is $1_A \otimes 1_A'$ and the idempotents are $e_i \otimes e_j'$ for all $i, j \in Q_0$. Hence the vertex set of the quiver

Q^e of A^e is $Q_0^e = \{(i,j') \mid i,j \in Q_0\}$. The arrows are $\alpha \otimes e'_j : (i,j') \to (k,j')$ and $e_j \otimes \alpha' : (j,k') \to (j,i')$ for $\alpha : i \to k \in Q_1$ and for all e_j.

Denote by I^e the ideal generated by $a \otimes e_i$ and $e_i \otimes a'$ for all generators a of I and all e_i. We get $A^e = KQ^e/I^e$. The indecomposable projective modules of A^e are $P(i,j') = A^e(e_i \otimes e'_j) = Ae_i \otimes (A^{op} * e'_j) = Ae_i \otimes (e_j A)^{op}$. The simple A^e-modules are $S(i,j') = P(i,j')/\mathrm{rad}P(i,j')$.

Then A is a left A^e-module with $(a \otimes b')x = axb$ for $x \in A$ and $a \otimes b' \in A^e$. This is equivalent to considering A as a bimodule over itself. We want to compute a projective resolution of A over A^e.

The standard resolution A classical resolution of A is the *standard resolution* (compare [5]). Let $S_i(A) = A^{\otimes_K i+2}$ denote the $(i+2)$-fold tensor product of A with itself. Then this is a projective A^e-module. Let $\delta_0 : A \otimes A \to A$ be the multiplication map $a_0 \otimes a_1 \mapsto a_0 a_1$. Define $\delta_i : S_i(A) \to S_{i-1}(A)$ for $i > 0$ by

$$\delta_i(a_0 \otimes \ldots \otimes a_{i+1}) = \sum_{j=0}^{i} (-1)^j a_0 \otimes \ldots \otimes a_j a_{j+1} \otimes \ldots \otimes a_{i+1}.$$

Then the complex

$$\cdots \xrightarrow{\delta_{i+1}} S_i(A) \xrightarrow{\delta_i} S_{i-1}(A) \xrightarrow{\delta_{i-1}} \cdots \xrightarrow{\delta_2} S_1(A) \xrightarrow{\delta_1} S_0(A) \xrightarrow{\delta_0} A \longrightarrow 0$$

is a projective resolution of A. In general it is not minimal.

The resolution of Happel There is a first description of a minimal projective resolution of A as an A^e-module, which is due to Happel.

Lemma 1.3. *[9, 1.5] Let*

$$\ldots R_n \longrightarrow R_{n-1} \longrightarrow \cdots \longrightarrow R_1 \longrightarrow R_0 \longrightarrow A \longrightarrow 0$$

be a minimal projective resolution of A over A^e. Then

$$R_n = \bigoplus_{i,j} P(i,j')^{\dim \mathrm{Ext}_A^n(S(j),S(i))}.$$

Remark 1.4. Let A be finite dimensional and let $S \in A$-mod be a simple A-module. Let (P_i, d_i) be a minimal projective resolution of S. Then we will often use the following well known fact together with Happel's Lemma:

$$\mathrm{Ext}_A^m(S, S') = \mathrm{Hom}_A(P_m, S') \tag{1.1}$$

for all $m \geq 0$ and all simple A-modules S'.

Definition 1.5. A frequently used application of such a resolution of A is the *Hochschild cohomology* of A. It is defined as

$$H^i(A) = \mathrm{Ext}_{A^e}^i(A, A).$$

Orderings and tip To understand the two concepts of resolutions in section 1.2 and 1.3 we need some further theory. Recall that B denotes the set of all paths in KQ. The first step is to define an ordering on B. There are several terms of a "good" ordering. In [7] they required a total order on B that satisfies the minimum condition. Bardzell [4] used the length-lexicographic order (see below) and called it admissible. He did not use it for the definition of his resolution but for the proof of its minimality. Anick and Green [2] used the following definition:

Definition 1.6. By an *ordering* on B we mean a partial order which becomes a total order when restricted to each $B^i = \{\beta \in B \mid s(\beta) = i\}$. An ordering $<$ is *suitable* if and only if the following three axioms are satisfied:

(1) Each B^i is well-ordered, with minimal element v_i.

(2) Whenever $\beta_1 > \beta_2$ and $\beta_1\gamma$ and $\beta_2\gamma$ are defined, then $\beta_1\gamma > \beta_2\gamma$.

(3) Whenever $\gamma_1 > \gamma_2$ and $\beta\gamma_1$ and $\beta\gamma_2$ are defined, then $\beta\gamma_1 > \beta\gamma_2$.

In every case the following order satisfies the conditions. Suppose $e_1 < e_2 < \ldots < e_n < \alpha_1 < \ldots < \alpha_n$ is a total order on $Q_0 \cup Q_1$. Then let $<$ be the *length-lexicographic order* on B. This means $v < w$ for two paths $v, w \in B$ if either $l(v) < l(w)$ or $l(v) = l(w)$ and $v < w$ in a lexicographic sense by the ordering of the "characters" above. From now on this is our order on B.

Definition 1.7. Let $x = \sum_{i=1}^{l} k_i w_i$, $w_i \in B$, be an element of KQ. The set $\text{supp}(x) = \{w_1, \ldots, w_l\}$ is called the *support* of x. We define $\text{tip}(x)$ to be the largest path under the admissible order in the support of x and $\text{Tip}(I)$ as the set of paths which are tips of elements of I. Finally set $\text{NonTip}(I) = B \setminus \text{Tip}(I)$.

There is a K-vector space decomposition $KQ = I \oplus \text{span}(\text{NonTip}(I))$. Every element $x \in KQ \setminus \{0\}$ can be written as $x = x_I + x_N$ where $x_I \in I$ and $x_N \in \text{span}(\text{NonTip}(I))$. Now consider the ideal $\langle \text{Tip}(I) \rangle$. It is monomial and one can choose a minimal set of generators $p_1, \ldots, p_m$ with $p_i = (p_i)_I + (p_i)_N$. Set $\text{Minsharp}_<(I) = \{(p_1)_I, \ldots, (p_m)_I\}$. This set was used by Bardzell. With the following theorem we can replace it by a minimal set of generators of I if it is a monomial ideal.

Theorem 1.8. *[7, Theorem 13] Let A be monomial, let S be a subset of elements in KQ which generate the ideal I. Suppose that*

(a) *the elements of S are paths with coefficient $1 \in K$, and*

(b) *no subpath of an element of S lies in S.*

Then $S = \text{Minsharp}_<(I)$.

From now on let $I = \langle S \rangle$, S with the above properties is the set of generating relations.

4

1.2   The resolution for monomial algebras

In this section we will define a minimal resolution for monomial algebras. The projective modules are known by Happel (Lemma 1.3), the construction of associated sequences (see below) goes back to [2] and [8]. With this technique, finally Bardzell [4] gave the morphisms.

Throughout $A = KQ/I$ is a finite dimensional monomial algebra and I is generated by a minimal set of paths S.

Definition 1.9. Let T be a directed path in Q. Let $M' \subset S$ be the subset of all those elements in S that are subpaths of T. Define a particular ordering on the vertices along T where $i < j$ for $i, j \in Q_0$ if there is a subpath of T starting in i and ending in j. Let $p_i \in M'$. We define the *associated sequence of paths* corresponding to p_i inductively as follows: Let $r_2 \in M'$ be the path (if it exists) in M' such that $s(p_i) < s(r_2) < e(p_i)$ and $s(r_2)$ is minimal with respect to this double inequality. Now assume $r_1, r_2, \ldots, r_j$ have been constructed, where $r_1 = p_i$. Let

$$L_{j+1} = \{r \in M' \mid e(r_{j-1}) \leq s(r) < e(r_j)\}.$$

If $L_{j+1} \neq \emptyset$, let r_{j+1} be such that $s(r_{j+1})$ is minimal with respect to $r_{j+1} \in L_{j+1}$. If $L_{j+1} = \emptyset$, the sequence stops at this point.

This is called the left construction for associated paths. There is a dual way to define the right construction:

Definition 1.10. Assume the same hypotheses as in the previous definition. Let $r_2 \in M'$ be the path (if it exists) in M' such that $s(r_2) < s(p_i) < e(r_2)$ and $e(r_2)$ is maximal with respect to this double inequality. Now assume $r_1, r_2, \ldots, r_j$ have been constructed. Let

$$R_{j+1} = \{r \in M' : s(r_j) < e(r) \leq s(r_{j-1})\}.$$

If $R_{j+1} \neq \emptyset$, let r_{j+1} be such that $e(r_{j+1})$ is maximal with respect to $r_{j+1} \in R_{j+1}$. If $R_{j+1} = \emptyset$, the sequence stops.

Remark 1.11. The ordering in the above definitions is not an ordering on the set of paths. Consider for Example T as a path repeatedly going through one cycle. Then we need an ordering which allows eventually "$n < n$" for some vertex n on the cycle. We can simply renumber the vertices T runs through starting with $s(T) = 1$ and increase with every arrow by one. Now everything works (see examples 1.14, 1.15) although one vertex can have many different numbers.

There is another way to define the sequences without any ordering. The condition for r_2 in the left-construction is that the first few arrows of the path r_2 coincide with the last few arrows of p_1, which means that both paths overlap. Choose r_2 with the largest overlapping with p_1. The condition for r_{j+1} is that r_{j+1} and

5

r_{j-1} don't overlap and that r_{j+1} and r_j overlap. Choose the r_{j+1} with the largest overlapping with r_j. The right-construction can be formulated in a dual way.

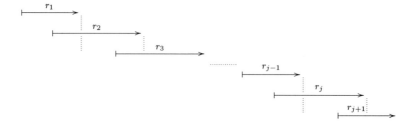

The projective modules of the resolution   Let $S = \{p_1, \ldots, p_k\}$. Let n be an integer and let $(r_1, \ldots, r_{n-1})$ be the tuple of the first $n-1$ paths of an associated sequence corresponding to $r_1 = p_i$ by the left construction. Let p_i^n be the path along all the paths of the sequence, which means along the directed path T, with starting point $s(r_1)$ and ending point $e(r_{n-1})$. There can be many paths along which one may construct another sequence for p_i. Let $AP_i(n)$ be the set of all these p_i^n associated to p_i. Set $AP(0) = Q_0$ and $AP(1) = Q_1$. Define for $n \geq 2$

$$AP(n) = \bigcup_{i=1}^{k} AP_i(n).$$

Then $AP(2) = S$. There is a dual way to go on with the right construction and one gets $AP(n)^{op}$ with $AP(n) = AP(n)^{op}$ ([4, 3.1]).

These sets $AP(n)$ determine the projective modules of the resolution:

$$P_n = \bigoplus_{p^n \in AP(n)} P(e(p^n), s(p^n)')$$

for all $n \geq 0$.

The morphisms of the resolution   The morphisms $\Phi_n : P_n \to P_{n-1}$, $n \geq 1$, of the resolution result from the divisor of a path p_i^n divided by a subpath p_i^{n-1}. More precisely let $p^n \in AP(n)$ and define

$$\mathrm{Sub}(p^n) = \{p^{n-1} \in AP(n-1) \mid p^{n-1} \text{ divides } p^n\}.$$

Bardzell showed that there are at least two paths in $\mathrm{Sub}(p^n)$, namely p_0^{n-1} with $s(p_0^{n-1}) = s(p^n)$ and p_1^{n-1} with $e(p_1^{n-1}) = e(p^n)$. It is possible that these paths coincide (see Example 2.5). He also showed that if n is odd, $\mathrm{Sub}(p^n) = \{p_0^{n-1}, p_1^{n-1}\}$ and there are paths $L, R \in KQ$ with $p^n = Lp_0^{n-1} = p_1^{n-1}R$ ([4, 3.3]). Set

$$\Phi_n(e(p^n) \otimes s(p^n)') = L \otimes s(p^n)' - e(p^n) \otimes R'.$$

6

For n even he showed that $\text{Sub}(p^n) = \{p_0^{n-1}, p_1^{n-1}, \dots, p_l^{n-1}\}$. Moreover, there exist paths θ_i and μ_i such that $p^n = \theta_i p_i^{n-1} \mu_i$ for all i. Set

$$\Phi_n(e(p^n) \otimes s(p^n)') = \sum_{i=1}^{l} \theta_i \otimes \mu_i'.$$

None of these paths L, R, θ_i, μ_i is divisible by an element of S ([4, 3.4]).

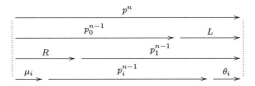

Theorem 1.12. *[4, 4.1] Let Q be a finite quiver. Suppose $A = KQ/I$ is a monomial algebra and I is admissible. Let P_n and Φ_n be as above and let π be the multiplication map. Then the sequence*

$$\dots \to P_{n+1} \xrightarrow{\Phi_{n+1}} P_n \xrightarrow{\Phi_n} \dots \xrightarrow{\Phi_2} P_1 \xrightarrow{\Phi_1} P_0 \xrightarrow{\pi} A \to 0$$

is a minimal projective resolution of A as a left A^e-module.

Remark 1.13. It is easy to see that the length of the resolution depends on the length of the paths that can be built by overlapping elements of S.

If A is hereditary ($I = 0$), the modules P_n are zero for all $n \geq 2$. This follows from the definition with $AP(2) = \emptyset$.

If the path algebra KQ is finite dimensional, the basis B of KQ as a K-vector space is finite. Then there exists a path of maximal length. Because the length of the paths $p^n \in AP(n)$ grow with bigger n, the resolution of A is finite.

The following three Examples explain the above definitions. They give also a first idea of periodicity.

Example 1.14. Let $A = KQ/KQ^{+2}$ be the algebra with radical square zero and the quiver

$$Q = 1 \xrightarrow{\alpha_1} 2$$
$$\alpha_3 \diagdown \quad \diagup \alpha_2$$
$$3$$

Then the ideal is generated by $S = \{p_1 = \alpha_2\alpha_1, \ p_2 = \alpha_3\alpha_2, \ p_3 = \alpha_1\alpha_3\}$. Let T be the directed path along the cycle of infinite length. Thus M' in definition 1.9 is equal to S. In this case we see that the ordering on the set of paths is not useful. Without loss of generality say $s(T) = 1$ and renumber the vertices on T by increasing by one for each arrow. We want to construct the associated sequence of paths for $r_1 = p_1$.

$$1 = s(p_1) < s(r_2) < e(p_1) = 3$$

Hence $r_2 = p_2$ with $s(p_2) = 2$. The next step is to build the set $L_3 = \{r \in M' \mid 3 = e(r_1) \le s(r) < e(r_2) = 4(= 1)\}$ and choose the element with minimal starting point which is $r_3 = p_3$.

$$
\begin{aligned}
p_1 &= \xrightarrow{\alpha_1} \xrightarrow{\alpha_2} \\
r_2 &= \xrightarrow{\alpha_2} \xrightarrow{\alpha_3} \\
r_3 &= \xrightarrow{\alpha_3} \xrightarrow{\alpha_1} \\
r_4 &= \xrightarrow{\alpha_1} \xrightarrow{\alpha_2} \\
p_1^5 &= \xrightarrow{\alpha_1} \xrightarrow{\alpha_2} \xrightarrow{\alpha_3} \xrightarrow{\alpha_1} \xrightarrow{\alpha_2}
\end{aligned}
$$

The sequence becomes periodic with a period of length 3 as one can see in the diagram. The associated sequences for p_2 and p_3 can be constructed analogously.

In this example there is no difference between the construction of r_2 and r_j for $j > 2$ (compare the definition) because all elements of S have length 2 and there is only one vertex in the middle of each path. Set $P_0 = P(1, 1') \oplus P(2, 2') \oplus P(3, 3')$, $P_1 = P(2, 1') \oplus P(3, 2') \oplus P(1, 3')$, and $P_2 = P(3, 1') \oplus P(1, 2') \oplus P(2, 3')$. We get the following resolution

$$
\ldots \to P_2 \to P_1 \to P_0 \to P_2 \to P_1 \to P_0 \to A \to 0
$$

of A as an A^e-module. The maps are also periodic and easy to compute.

Example 1.15. Let Q be as in Example 1.14 and let S be the set $\{p_1 = \alpha_3\alpha_2\alpha_1, p_2 = \alpha_1\alpha_3\alpha_2, p_3 = \alpha_2\alpha_1\alpha_3\}$. Then the sequence for p_1 looks as follows:

$$
\begin{aligned}
p_1 &= \xrightarrow{\alpha_1} \xrightarrow{\alpha_2} \xrightarrow{\alpha_3} \\
r_2 &= \xrightarrow{\alpha_2} \xrightarrow{\alpha_3} \xrightarrow{\alpha_1} \\
r_3 &= \xrightarrow{\alpha_1} \xrightarrow{\alpha_2} \xrightarrow{\alpha_3} \\
r_4 &= \xrightarrow{\alpha_2} \xrightarrow{\alpha_3} \xrightarrow{\alpha_1} \\
p_1^5 &= \xrightarrow{\alpha_1} \xrightarrow{\alpha_2} \xrightarrow{\alpha_3} \xrightarrow{\alpha_1} \xrightarrow{\alpha_2} \xrightarrow{\alpha_3} \xrightarrow{\alpha_1}
\end{aligned}
$$

Here we can see the difference between the construction in the first and the later steps. The element r_3 cannot be chosen to start with α_3 because it must not overlap with $p_1 = r_1$.

Example 1.16. Let Q be the following quiver

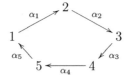

8

Let I be generated by the set $S = \{p_1 = \alpha_3\alpha_2\alpha_1,\ p_2 = \alpha_4\alpha_3\alpha_2,\ p_3 = \alpha_5\alpha_4\alpha_3,\ p_4 = \alpha_2\alpha_1\alpha_5\alpha_4\}$. For the algebra $A = KQ/I$ we compute the associated sequence for $p_2 = \alpha_4\alpha_3\alpha_2$ as follows:

$$r_1 = \xrightarrow{\alpha_2} \xrightarrow{\alpha_3} \xrightarrow{\alpha_4}$$
$$r_2 = \qquad\ \xrightarrow{\alpha_3} \xrightarrow{\alpha_4} \xrightarrow{\alpha_5}$$
$$p_3 = \xrightarrow{\alpha_2} \xrightarrow{\alpha_3} \xrightarrow{\alpha_4} \xrightarrow{\alpha_5}$$

The next set L_3 is empty and the sequence stops. The sequence for $p_1 = \alpha_3\alpha_2\alpha_1$ is infinite and looks as follows:

This sequence becomes periodic after the second element; $r_3 = r_6 = r_9 = \ldots$. We do not have periodicity right from the start because the second element starts at the vertex 2 by $s(r_1) < s(r_2) < e(r_1)$ for the first construction step. The later steps refer to $e(r_{n-1}) \leq s(r_{n+1}) < e(r_n)$. So r_5 must start at the vertex 3 just as all the following $r_5 = r_8 = \ldots$. The other two sequences can be computed analogously. This means we have three periodic sequences (after a certain point) and one finite sequence. Hence the resolution of the algebra is infinite and periodic after a certain point.

The projective resolution of the simple A-modules Now we can give an explicit description of a minimal projective resolution of the simple A-modules in terms of Bardzell's resolution.

Proposition 1.17. *Let $i \in Q_0$ be a vertex in the quiver and let $AP(n)_i = \{p^n \in AP(n) \mid s(p^n) = i\}$.*

(1) Then the set $\mathrm{Sub}(p^n) \cap AP(n-1)_i$ consists only of the element p^{n-1} with $p^n = Lp^{n-1}$.

(2) Define

- *$P_n^i = \bigoplus_{p^n \in AP(n)_i} P(e(p^n))$ and*
- *$\Phi_n^i : P_n^i \to P_{n-1}^i$ by $\Phi_n^i(e(p^n)) = L$ for all $n \geq 1$ and*
- *let $\Phi_0^i : P_0^i \to S(i)$ be the projective cover with $P_0^i = P(i)$.*

Then (P_n^i, Φ_n^i) is a minimal projective resolution of the simple A-module $S(i)$.

Proof. (1) Every path $q^{n-1} \in AP(n-1)$ that starts in i and is a subpath of p^n must be associated to the same sequence because of the uniqueness of the construction. So it is equal to p^{n-1} and there is a path L such that $p^n = Lp^{n-1}$.

(2) First we proof that the sequence is exact. Consider

$$\Phi^i_{n-1} \circ \Phi^i_n(e(p^n)) = \Phi^i_{n-1}(L) = L\Phi^i_{n-1}(e(p^{n-1})) = LL'$$

with $p^{n-1} = L'p^{n-2}$ for all $n > 1$. Let $(r_1, r_2, \ldots, r_{n-2}, r_{n-1})$ be the associated sequence of p^n where $r_l \in S$ for all l. By construction we have $e(p_{n-2}) = e(r_{n-3}) \leq s(r_{n-1}) < e(r_{n-2}) = e(p_{n-1})$. Hence r_{n-1} is a subpath of LL' and $\Phi^i_{n-1} \circ \Phi^i_n = 0$ for all $n \geq 2$.

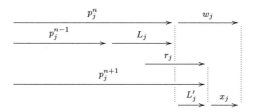

For $n = 1$ we have to consider $\alpha \in AP(1)_i$ where $\alpha : i \to j$ is an arrow in Q. Then $\Phi^i_0 \circ \Phi^i_1(e(\alpha)) = \Phi^i_0(\alpha) = 0$ because $\alpha \in \mathrm{rad}\, P(i) = \ker \Phi^i_0$.

Let now $w = \sum_j w_j \in \ker \Phi^i_n$ with $w_j \in P(e(p^n_j))$ a path. Thus $\Phi^i_n(w) = \sum_j w_j L_j = 0$. This means for every j there exists an $r_j \in S$ that is a subpath of $w_j L_j$. Choose the one with minimal starting point. Then this will be the next path in the sequence of p^{n+1}_j. Because L_j cannot be divided by any element of S, we have an L'_j with $p^{n+1}_j = L'_j p^n_j$. Let x_j denote the path from $e(L'_j)$ to $e(w_j)$.

$$\begin{array}{c}
\underrightarrow{\hspace{2cm} p^n_j \hspace{2cm}} \quad \underrightarrow{\; w_j \;} \\
\underrightarrow{\; p^{n-1}_j \;} \quad \underrightarrow{\; L_j \;} \\
\qquad \qquad \underrightarrow{\; r_j \;} \\
\underrightarrow{\hspace{3cm} p^{n+1}_j \hspace{3cm}} \\
\qquad \qquad \underrightarrow{\; L'_j \;} \quad \underrightarrow{\; x_j \;}
\end{array}$$

Then $\Phi^i_{n+1}(\sum_j x_j) = \sum_j x_j \Phi^i_{n+1}(e(p^{n+1}_j)) = \sum_j x_j L'_j = \sum_j w_j = w$ and $w \in \mathrm{Im}\, \Phi^i_{n+1}$.

Finally we show that the resolution is minimal. This means the kernel of every map $P^i_n \to \mathrm{Im}\, \Phi^i_n$ is a part of the radical of the projective module P^i_n. Let $w = \sum_j w_j \in P^i_n$ be in the kernel of Φ^i_n, thus $\sum_j w_j L_j = 0$. Because $L_j \notin I$, w_j must be a path of positive length such that $w_j L_j \in I$. Thus $w_j \in \mathrm{rad}\, P^i_n$ for all j. $\qquad \square$

1.3 The resolution of Anick and Green

The following resolution due to Anick and Green [2] is a projective resolution for simple modules of any basic and finite dimensional algebra. It is based on the results of Anick in [1]. Here we do not get a resolution of A itself directly. But we can use this resolution to determine the modules of a minimal resolution of A using Happel's result (Lemma 1.3). In particular this works in the non-monomial case.

From now on let A be an associative K-algebra, which does not need to be monomial but split basic. Again $A = KQ/I$, Q is a finite quiver and I is an admissible ideal of KQ. Let B be the set of paths in Q which is a K-basis of KQ. Let $R = A/\mathrm{rad} A$. Moreover we need a suitable ordering $<$ on B (compare Definition 1.6). Let $f : KQ \to A$ be the surjection with kernel I. The first step is to construct a subset M of B such that $C := f(M)$ is a vector space basis for A.

Lemma 1.18. *[2, 2.2] Let $<$ be a suitable ordering on B and let $f : KQ \to A$ be an R-module surjection. Put*

$$M = \{\beta \in B \mid f(\beta) \notin \mathrm{span}(f(\gamma) \mid \gamma < \beta)\}.$$

Then $C = \{f(\beta) \mid \beta \in M\}$ is a K-basis for A. Furthermore, M has the property that $\beta \in M$ and $\gamma \in M$ whenever $\beta\gamma \in M$.

In this case we can use the length-lexicographic order ([2, 2.3]) for a local well order on $Q_0 \cup Q_1$. Define the *obstruction set* Q_2 for A relative to $<$ and f as $Q_2 = \{\beta \in B \setminus M \mid \text{whenever } \beta = \beta_1\beta_2 \text{ with } \beta_1, \beta_2 \in B \setminus Q_0, \text{ then } \beta_1 \in M \text{ and } \beta_2 \in M\}$.

Lemma 1.19. *[2, 2.4] Let $\gamma \in B$. Then $\gamma \in B \setminus M$ if and only if there exists $\beta \in Q_2$ which is a subpath of γ.*

Now we can write $M = \{\beta \in B \mid \text{no subpath of } \beta \text{ lies in } Q_2\}$.

Definition 1.20. Recursively define the set of *m-chains* on the set Q_2, $m \geq -1$, as follows. For $m = -1, 0, 1$, an m-chain is defined to be an element of Q_{m+1}. Now let $m \geq 1$ and suppose that the set Q_{m+1} of m-chains has been defined. Supposing also that each $\beta \in Q_{m+1}$ has a unique right factor β_1 which is an $(m-1)$-chain, define the set Q_{m+2} of $(m+1)$-chains to be the set of words $\gamma \in B$ such that

(1) $\gamma = \gamma_2\gamma_1$ for some $\gamma_1 \in Q_{m+1}$ and $\gamma_2 \in M \setminus Q_0$.
(2) If $\gamma = \gamma_2\gamma_1$ with $\gamma_1 \in Q_{m+1}$ and we factor γ_1 as $\gamma_1 = \beta_2\beta_1$ with $\beta_1 \in Q_m$, then $\gamma_2\beta_2 \in B \setminus M$.
(3) γ does not equal $\gamma''\gamma'$, where $\gamma'' \in B \setminus Q_0$ and γ' also satisfies (1) and (2) above.

The projective modules of the resolution   These sets of m-chains are the base for the projective modules in the Anick/Green-resolution. Put $Q_m^i = Q_m \cap B^i$ and let KQ_m denote a vector space with basis Q_m. Thus $KQ_m^i = (KQ_m)e_i$ and

$$K_m := A \otimes_R KQ_m^i \simeq \bigoplus_{\beta \in Q_m^i} Ae_{e(\beta)}$$

11

is a projective left A-module. This will be the m-th module in the resolution of the simple module S_i.

The morphisms of the resolution Define for any $\beta \in B^i$ the R-submodule W_m^β of $A \otimes_R KQ_m^i$ to be

$$W_m^\beta = \text{span}\{f(\beta_2) \otimes \beta_1 \mid \beta_2 \otimes \beta_1 \in M \otimes Q_m^i \text{ and } \beta_2\beta_1 < \beta\}.$$

Every $\beta \in B^i$ has at most one factorization $\beta = \beta_1\beta_2$ where $\beta_1 \in M$ and $\beta_2 \in Q_m^i$. If such a factorization exists, let $\beta[m] = \beta_1 \otimes \beta_2$. Then $w \equiv (c_1)\beta[m] \pmod{W_m^\beta}$, for $c_1 \in k$ and $w \in A \otimes_R KQ_m^i$, is equivalent to the assertion that w may be expressed as

$$w = c_1\beta_1 \otimes \gamma_1 + c_2\beta_2 \otimes \gamma_2 + \ldots + c_s\beta_s \otimes \gamma_s$$

where $\beta_1\gamma_1 = \beta$ and $\beta > \beta_j\gamma_j$ for $1 < j \leq s$. The morphisms $\delta_m : K_m \to K_{m-1}$ of the resolution of S_i are defined by

$$\delta_m(v_{e(\beta)} \otimes \beta) \equiv \beta[m-1] \pmod{W_{m-1}^\beta}$$

for $\beta \in Q_m^i$ and $m \geq 1$.

Theorem 1.21. *[2, 2.7] Let A be a basic K-algebra, let $f : KQ \to A$ be a surjection with $\ker f \subseteq \text{rad}KQ$ and let $e_1, \ldots, e_n$ be the idempotents in A corresponding to the vertices $1, \ldots, n$ of Q. Choose any suitable ordering on the set B of paths on Q and let Q_2 be the corresponding obstruction set determined via Lemma 1.18. Fix i, $1 \leq i \leq n$, and let K_m denote the projective left A-module $A \otimes_R KQ_m^i$. Then there is an exact sequence of left A-modules*

$$\ldots \to K_3 \xrightarrow{\delta_3} K_2 \xrightarrow{\delta_2} K_1 \xrightarrow{\delta_1} K_0 (= Ae_i) \xrightarrow{\delta_0} Ae_i/\text{rad}Ae_i \to 0 \qquad (1.2)$$

in which, for $\beta \in Q_m^i$ and $m \geq 1$,

$$\delta_m(v_{e(\beta)} \otimes \beta) \equiv \beta[m-1] \pmod{W_{m-1}^\beta}.$$

Corollary 1.22. *[2, 2.9] Let $\rho \in B \setminus (Q_1 \cup Q_0)$ be any set of zero-relations on KQ which is minimal. That is, ρ is a set of paths of length two or greater such that no proper subpath of any $\beta \in \rho$ belongs to ρ. Let I be the two-sided ideal of KQ which ρ generates and let $A = KQ/I$. Then any admissible ordering leads to the same obstruction set, namely $Q_2 = \rho$. Furthermore, the resolution (1.2) is minimal in the sense that each $\delta_m \otimes (\text{id})_R : K_m \otimes_A R \to K_{m-1} \otimes_A R$ is zero. In particular, $\text{Tor}_m^A(Ae_i/\text{rad}Ae_i, R) \simeq \text{span}_k(Q_m^i)$.*

By this corollary we know that for a monomial algebra the Anick/Green-resolution is minimal. So it coincides with the resolution of Bardzell in the version for simple modules (Proposition 1.17) up to isomorphism. Bardzell's resolution is easier to compute but the Anick/Green-resolution is also defined in the non-monomial case.

Example 1.23. Let Q be the quiver

$$Q = \quad \begin{array}{c} \overset{2}{} \\ {}^{\alpha_1}\nearrow \quad \searrow{}^{\alpha_2} \\ 1 \xrightarrow[]{\beta} 3 \\ {}_{\alpha_4}\nwarrow \quad \swarrow{}_{\alpha_3} \\ 4 \end{array}$$

Let I be an ideal of KQ generated by the set of relations

$$S = \{\alpha_4\alpha_3,\ \alpha_1\alpha_4,\ \alpha_3\beta - \alpha_3\alpha_2\alpha_1\}$$

and let $A = KQ/I$. The projective modules of A in terms of their composition factors are:

$$P(1) = \begin{pmatrix} & 1 & \\ 2 & & 3 \\ 3 & & \\ & 4 & \end{pmatrix}, \quad P(2) = \begin{pmatrix} 2 \\ 3 \\ 4 \end{pmatrix}, \quad P(3) = \begin{pmatrix} 3 \\ 4 \end{pmatrix}, \quad P(4) = \begin{pmatrix} 4 \\ 1 \\ 3 \end{pmatrix}.$$

Then the resolutions of Anick/Green are not minimal. By the length-lexicographic order we get $Q_2 = \{\alpha_4\alpha_3,\ \alpha_1\alpha_4,\ \alpha_3\alpha_2\alpha_1\}$ and

$$M = \{e_1,\ e_2,\ e_3,\ e_4,\ \alpha_1,\ \alpha_2,\ \alpha_3,\ \alpha_4,\ \beta,\ \alpha_2\alpha_1,\ \alpha_3\alpha_2,\ \beta\alpha_4,\ \alpha_3\beta\}.$$

The sets of the m-chains that start in the vertex 1 are $Q_0^1 = \{e_1\}$, $Q_1^1 = \{\alpha_1, \beta\}$, $Q_2^1 = \{\alpha_3\alpha_2\alpha_1\}$, and

$$
\begin{aligned}
Q_3^1 &= \{\alpha_4\alpha_3\alpha_2\alpha_1\} \\
Q_4^1 &= \{\alpha_1\alpha_4\alpha_3\alpha_2\alpha_1,\ \alpha_3\beta\alpha_4\alpha_3\alpha_2\alpha_1\} \\
Q_5^1 &= \{\alpha_3\alpha_2\alpha_1\alpha_4\alpha_3\alpha_2\alpha_1,\ \alpha_4\alpha_3\beta\alpha_4\alpha_3\alpha_2\alpha_1\} \\
Q_6^1 &= \{\alpha_4\alpha_3\alpha_2\alpha_1\alpha_4\alpha_3\alpha_2\alpha_1,\ \alpha_3\beta\alpha_4\alpha_3\beta\alpha_4\alpha_3\alpha_2\alpha_1,\ \alpha_1\alpha_4\alpha_3\beta\alpha_4\alpha_3\alpha_2\alpha_1\} \\
Q_7^1 &= \{\alpha_3\beta\alpha_4\alpha_3\alpha_2\alpha_1\alpha_4\alpha_3\alpha_2\alpha_1,\ \alpha_1\alpha_4\alpha_3\alpha_2\alpha_1\alpha_4\alpha_3\alpha_2\alpha_1, \\
&\qquad \alpha_4\alpha_3\beta\alpha_4\alpha_3\beta\alpha_4\alpha_3\alpha_2\alpha_1,\ \alpha_2\alpha_1\alpha_4\alpha_3\beta\alpha_4\alpha_3\alpha_2\alpha_1\} \\
Q_8^1 &= \{\alpha_4\alpha_3\beta\alpha_4\alpha_3\alpha_2\alpha_1\alpha_4\alpha_3\alpha_2\alpha_1,\ \alpha_2\alpha_1\alpha_4\alpha_3\alpha_2\alpha_1\alpha_4\alpha_3\alpha_2\alpha_1, \\
&\qquad \alpha_1\alpha_4\alpha_3\beta\alpha_4\alpha_3\beta\alpha_4\alpha_3\alpha_2\alpha_1,\ \alpha_3\beta\alpha_4\alpha_3\beta\alpha_4\alpha_3\beta\alpha_4\alpha_3\alpha_2\alpha_1, \\
&\qquad \alpha_3\alpha_2\alpha_1\alpha_4\alpha_3\beta\alpha_4\alpha_3\alpha_2\alpha_1\}
\end{aligned}
$$

and so on. Thus the resolution of Anick/Green of the simple A-module $S(1)$ is infinite and not periodic:

$$
\begin{aligned}
0 \ &\leftarrow S(1) \leftarrow P(1) \leftarrow P(2) \oplus P(3) \leftarrow P(4) \leftarrow P(1) \leftarrow \\
&\leftarrow P(2) \oplus P(4) \leftarrow P(4) \oplus P(1) \leftarrow P(1) \oplus P(2) \oplus P(4) \leftarrow \\
&\leftarrow P(1) \oplus P(2) \oplus P(4)^2 \leftarrow P(1)^2 \oplus P(2) \oplus P(4)^2 \leftarrow \dots
\end{aligned}
$$

13

It is not minimal as one can see in terms of the composition factors (compare the kernels):

$$0 \;\leftarrow\; (1) \;\leftarrow\; \left(\begin{smallmatrix} & 1 & \\ 2 & & 3 \\ 3 & & \\ & 4 & \end{smallmatrix}\right) \;\leftarrow\; \left(\begin{smallmatrix}2\\3\\4\end{smallmatrix}\right) \oplus \left(\begin{smallmatrix}3\\4\end{smallmatrix}\right) \;\leftarrow\; \left(\begin{smallmatrix}4\\1\\3\end{smallmatrix}\right) \;\leftarrow\; \left(\begin{smallmatrix} & 1 & \\ 2 & & 3 \\ & 4 & \end{smallmatrix}\right) \;\leftarrow$$

$$\leftarrow\; \left(\begin{smallmatrix}2\\3\\4\end{smallmatrix}\right) \oplus \left(\begin{smallmatrix}4\\1\\3\end{smallmatrix}\right) \;\leftarrow\; \left(\begin{smallmatrix}4\\1\\3\end{smallmatrix}\right) \oplus \left(\begin{smallmatrix} & 1 & \\ 2 & & 3 \\ & 4 & \end{smallmatrix}\right) \;\leftarrow\; \left(\begin{smallmatrix} & 1 & \\ 2 & & 3 \\ & 4 & \end{smallmatrix}\right) \oplus \left(\begin{smallmatrix}2\\3\\4\end{smallmatrix}\right) \oplus \left(\begin{smallmatrix}4\\1\\3\end{smallmatrix}\right) \;\leftarrow$$

$$\leftarrow\; \left(\begin{smallmatrix} & 1 & \\ 2 & & 3 \\ & 4 & \end{smallmatrix}\right) \oplus \left(\begin{smallmatrix}2\\3\\4\end{smallmatrix}\right) \oplus \left(\begin{smallmatrix}4\\1\\3\end{smallmatrix}\right)^2 \;\leftarrow\; \left(\begin{smallmatrix} & 1 & \\ 2 & & 3 \\ & 4 & \end{smallmatrix}\right)^2 \oplus \left(\begin{smallmatrix}2\\3\\4\end{smallmatrix}\right) \oplus \left(\begin{smallmatrix}4\\1\\3\end{smallmatrix}\right)^2 \;\leftarrow\; \ldots$$

Actually, the minimal projective resolution of $S(1)$ is finite:

$$0 \leftarrow S(1) \leftarrow P(1) \leftarrow P(2) \oplus P(3) \leftarrow P(4) \leftarrow P(1) \leftarrow P(2) \leftarrow 0.$$

In terms of the composition factors:

$$0 \leftarrow (1) \leftarrow \left(\begin{smallmatrix} & 1 & \\ 2 & & 3 \\ 3 & & \\ & 4 & \end{smallmatrix}\right) \leftarrow \left(\begin{smallmatrix}2\\3\\4\end{smallmatrix}\right) \oplus \left(\begin{smallmatrix}3\\4\end{smallmatrix}\right) \leftarrow \left(\begin{smallmatrix}4\\1\\3\end{smallmatrix}\right) \leftarrow \left(\begin{smallmatrix} & 1 & \\ 2 & & 3 \\ & 4 & \end{smallmatrix}\right) \leftarrow \left(\begin{smallmatrix}2\\3\\4\end{smallmatrix}\right) \leftarrow 0.$$

Chapter 2

The monomial case

With the resolution of Bardzell the monomial algebras are falling by the structure of their resolutions into three types: with finite resolution, with (eventually) periodic resolution and with infinite, non-periodic resolution. If the quiver of a monomial algebra $A = KQ/I$ and a minimal set S of generating relations of the ideal I are given, the structure of the resolution can be read off immediately (Theorem 2.9). In the monomial case the structure of the resolutions of the simple modules also determines the structure of the resolution of the algebra over its enveloping algebra (Corollary 2.10).

For the whole chapter let $A = KQ/I$ be a finite dimensional monomial K-algebra and let S be a minimal set of relations which generate I.

2.1 Quivers and periodicity

Notation. Let $w = \alpha_n \ldots \alpha_1$ be a path in a quiver Q. We call a vertex $e \in Q_0$ an *inner vertex of w* if e is the starting point of one of $\alpha_n, \ldots, \alpha_2$ and thus the ending point of one of $\alpha_{n-1}, \ldots, \alpha_1$.

Definition 2.1. An oriented cycle in Q: $i_1 \xrightarrow{\alpha_1} i_2 \xrightarrow{\alpha_2} \ldots \to i_n \xrightarrow{\alpha_n} i_1$ is *covered by* S if there is an associated sequence of paths $(r_1, \ldots, r_{m+1})$ where $p^m = \alpha_j \ldots \alpha_k x$ and $l(p^m) \geq n + 1 + l(x)$, x a path in Q. Additionally, the next element of the sequence must be some element $r_i = \alpha_s \ldots \alpha_t$ of the sequence. Hence there is an associated sequence $(r_1, \ldots, r_{i-1}, (r_i, \ldots, r_{m+1})^l)$ for every l.

Let $i_1 \xrightarrow{\alpha_1} i_2 \xrightarrow{\alpha_2} \ldots \to i_n \xrightarrow{\alpha_n} i_1$ and $j_1 \xrightarrow{\beta_1} j_2 \xrightarrow{\beta_2} \ldots \to j_{n'} \xrightarrow{\beta_{n'}} j_1$ be two cycles that are covered by S. They are called *connected by S* if there is a path $w : i_l \dashrightarrow j_k$ in Q such that for every $x, y \geq 0$ the path $(\beta_{k+1} \ldots \beta_k)^x w (\alpha_{l-1} \ldots \alpha_l)^y$ is a subpath of an element of $AP(m)$ for some m. In case two cycles share vertices or arrows, the common vertex or arrows can be used as a connection.

Lemma 2.2. *Let $A = KQ/I$ be a monomial finite dimensional K-algebra. The resolution of A is infinite if and only if there is at least one cycle in the quiver of A which is covered by S.*

Proof. Let the resolution of A be infinite. Then there is at least one infinite associated sequence of paths $(r_1, \ldots, r_m, \ldots)$. Because the quiver is finite and the set S is minimal, S is also finite. Hence there exists at least one element r_i of the sequence which repeats infinitely many times in the sequence. Let $\{i, i + l_1, i + l_2, \ldots\}$ be the indices such that $r_i = r_{i+l_j}$ and choose i minimal in this set. Consider the set $\{r_{i-1}, r_{i+l_1-1}, r_{i+l_2-1}, \ldots\}$ of the predecessors of r_i in the sequence. This is a subset of S and hence finite. So there are again indices such that $r_{i+l-1} = r_{i+l'-1}$, $l > l'$. Comparing the construction of the sequence, we get

$$e(r_{i+l-1}) \leq s(r_{i+l+1}) < e(r_{i+l})$$
$$e(r_{i+l'-1}) \leq s(r_{i+l'+1}) < e(r_{i+l'})$$

and finally $r_{i+l+k} = r_{i+l'+k}$ for every k. Hence the sequence is of the form $(r_1, \ldots, r_{i+l'-2}, (r_{i+l'-1}, \ldots, r_{i+l-2})^x)$ and we get a cycle covered by S.

If on the other hand there exists a cycle in the quiver like above, the assertion follows by the definition. $\qquad\square$

Definition 2.3. Let A be a finite dimensional K-algebra. Let the enveloping algebra be $A^e = A \otimes_K A^{op}$. Then A has an *eventually periodic resolution with period of length $N > 0$ periodic from level $p \geq 0$* if there is an infinite and minimal projective resolution (P^i, d^i) of A over A^e with $P^{i+N} \simeq P^i$ for all $i \geq p$ and $d^{i+N} = d^i$ for all $i \geq p + 1$:

$$0 \longleftarrow A \longleftarrow P^0 \xleftarrow{d^0} P^1 \xleftarrow{d^1} P^2 \longleftarrow \cdots$$

$$\cdots \longleftarrow P^p \xleftarrow{d^{p+1}} P^{p+1} \xleftarrow{d^{p+2}} P^{p+2} \longleftarrow \cdots \longleftarrow P^{p+N-1} \xleftarrow{d^p}$$
$$\xleftarrow{d^p} P^{p+N} \xleftarrow{d^{p+1}} P^{p+N+1} \xleftarrow{d^{p+2}} P^{p+N+2} \longleftarrow \cdots \longleftarrow P^{p+2N-1} \longleftarrow \cdots$$

(with vertical $\|$ identifications shown between the two rows)

If the resolution is finite, we have the special case $N = 0$ and $p = \mathrm{pdim}_{A^e} A$. For $p = 0$ this is the classical definition of a *periodic* algebra.

If the projective modules in the resolution of A are (eventually) periodic but we do not know anything about the morphisms, we call the resolution *almost eventually periodic*. We will see that in the monomial case both terms are equivalent (Lemma 2.7).

Equivalent to the definition is the fact that $\Omega^n_{A^e}(A) \simeq \Omega^{n+N}_{A^e}(A)$ for all $n \geq p+1$ with $\Omega^n_{A^e}(A) = \ker d^{n-1}$. A module $M \in A\text{-mod}$ is called *periodic* if there exists an $n > 0$ such that $\Omega^n_A(M) \simeq M$ holds. The module category $A\text{-mod}$ is called periodic if all finite dimensional A-modules without projective summands are periodic. This subcategory is called $A\text{-mod}_\mathcal{P}$.

The case $p = 0$ will be discussed in detail in section 2.3. We will often omit the word eventually. Except in section 2.3 we always mean the general, i.e., eventually periodic, case.

Remark 2.4. In the non-monomial case we use the Anick/Green-resolution and consider the resolutions of the simple modules. The above definitions of periodicity can be formulated for simple modules. With the Lemma of Happel (1.3), we get: The resolution of A is almost eventually periodic if and only if the resolutions of the simple modules of A are almost eventually periodic.

Example 2.5. Let $Q = 1 \overset{\curvearrowright}{} \beta$ be the quiver with one vertex and one loop β and let $A = KQ/KQ^{+3}$. Then the single associated sequence of paths to $p = \beta^3 = r_1$ is

The resolution $\ldots \overset{\Phi_3}{\to} P \overset{\Phi_2}{\to} P \overset{\Phi_1}{\to} P \to A \to 0$ of A is periodic from level zero with a period of length 2 because $AP(2n) = \{\beta^{3n}\}$ and $AP(2n+1) = \{\beta^{3n+1}\}$ for all $n \geq 0$. So the maps are

$$
\begin{aligned}
\Phi_{2n}(e_1 \otimes e_1) &= \beta^2 \otimes e_1 + \beta \otimes \beta + e_1 \otimes \beta^2 \quad \text{and} \\
\Phi_{2n+1}(e_1 \otimes e_1) &= \beta \otimes e_1 - e_1 \otimes \beta \quad \text{for all } n \geq 0.
\end{aligned}
$$

Here we have the case that there is just one element in any set $\text{Sub}(\beta^{3n+1}) = \{\beta^{3n}\}$ and $\text{Sub}(\beta^{3n}) = \{\beta^{3n-1}\}$ for all $n \geq 0$.

In Example 1.14 the algebra A has a periodic resolution periodic from level $p = 0$ and with a period of length $N = 3$.

In Example 1.16 the algebra has a periodic resolution from level $p = 4$ with a period of length $N = 3$.

Lemma 2.6. *If there exist two cycles in the quiver of A that are covered and connected by S, then A has an infinite resolution which is not (even almost) periodic.*

Proof. There are three cases:
  (a) There are two cycles in the quiver of A which are covered and connected by S.
  (b) There is one cycle in the quiver of A with multiple arrows which is covered by S. At one vertex with multiple arrows at least two arrows are inner arrows of paths $p^m \in AP(m)$.
  (c) There are two cycles in the quiver of A which are connected by S and which share vertices and arrows $i_1 \overset{\alpha_1}{\to} i_2 \ldots \overset{\alpha_n}{\to} i_n$ (at least the vertex i_1). There is a path $p^m \in AP(m)$ that has $\beta\alpha_n \ldots \alpha_1\gamma$ as a subpath where β is the next arrow in the first cycle and γ is the previous arrow in the second cycle.

By Lemma 2.2 the resolution of A is infinite in every case. We omit the proofs of (b) and (c) because they are special cases of (a), the connecting paths of the two cycles are just subpaths of the cycles.

Now assume (a). The number of elements of the set $AP(n)$ is the number of the paths that can be built by overlapping $n-1$ elements of S. We will show that the sets $AP(n)$ grow with bigger n. Thus the number of the indecomposable summands of the projective module at the grade n in the resolution grows with bigger n and the resolution cannot be periodic.

Let $p \in S$ be a path along the one of the two cycles where the connecting path starts. Then it is possible to choose a path along this cycle to build an associated sequence of paths $r = (r_1, \ldots r_n)$ with $r_1 = p$ which goes one time around the first cycle. And we can build a sequence $(r^l, r_1, \ldots, r_j)$ along the cycle up to the point where the connecting path starts. Let $r_v = (r_1^v, \ldots, r_w^v)$ be the connection of the two cycles by S. Let $r' = (r_1', \ldots, r_m')$ be the sequence one time along the second cycle starting at the end of the connecting path. Then we can build endless associated sequences of paths of the form $(r^l, r_1, \ldots, r_j, r_v, (r')^k)$ for all $l, k \geq 0$. Let now $x \in \mathbb{N}$ be bigger than $n + j + w$. Then there is a path p^x in $AP(x)$ which belongs to a sequence of the above form with $x - 1$ elements. How many different sequences of this form and length can exist?

The lengths of the connecting path and of the last part of the path along the first cycle are fixed. So consider $y = x - 1 - w - j = ln + km + z$ where $z \in \mathbb{N}$, $z < m$, counts the elements of the sequence that are at the end of the sequence and cover only a part of the second cycle. Then one can write y as a sum $y = l_{max}n + z_1$ for $k = 0$ until $y = n + k_{max}m + z_2$. In both cases k depends on the choice of l. Thus there are l_{max} possibilities to build a path of the form p^x with given x. The bigger x becomes the bigger becomes l_{max} and the bigger $AP(x)$ becomes. This means that the number of projective modules in the resolution of A at the level x becomes bigger and bigger. So the resolution cannot be periodic. □

Lemma 2.7. *The algebra A has an eventually periodic resolution periodic from some level $p \geq 0$ and with a period of positive length if and only if*

(a) there is at least one cycle in the quiver Q which is covered by S and

(b) no pair of cycles is connected by S.

Moreover, the resolution is almost eventually periodic if and only if it is eventually periodic.

Proof. ($\Rightarrow$) If A has an eventually periodic resolution, then the assertion follows by the Lemmas 2.2 and 2.6.

($\Leftarrow$) Now let Q be as described above. By Lemma 2.2 the resolution of A is infinite. If there is a cycle that is covered by S, one can build a sequence $(r_1, \ldots r_n)$ for every $p \in S$ that lies on the cycle. It has the property that the associated path p^{n+1} runs completely through the cycle at least one time and the next element in the sequence along the quiver must be r_1 again. This sequence can be iterated arbitrarily often such that the associated paths $p^{ln} \in AP(ln)$ contain a period. Which paths can one also be built with p in S? If one can get out of the cycle with S, there is no way to end in another cycle by hypothesis, the sequence will stop.

Let $((r_1, \ldots, r_n)^l, r_1, \ldots, r_m)$ be a sequence inside the cycle (it goes l-times completely along the cycle) with ending point $e(r_m)$. Let k_m be the number of paths that are covered by S and got out of the cycle at the vertex $e(r_m)$. Then there is the one path that goes further along the cycle, take r_{m+1}, or take another but finite path $((r_1, \ldots, r_n)^l, r_1, \ldots, r_m, q_1^{m,j})$, $j = 1, \ldots, k_m$. Let $(q_1^{m,j}, \ldots, q_{l_{m,j}}^{m,j})$ be these finite paths for $j = 1, \ldots, k_m$. So the number of possible paths stays the same with $k_m + 1$ for every l.

How many paths are in S that start with the same p as above? Let $p = r_1$, then there are $1 + k_1$ sequences of the length 2 and $a_3 = 1 + k_1$ possible paths p^3.

In the next step there is again a path along the cycle (r_1, r_2, r_3). Then there is the continued paths $(r_1, q_1^{1,j}, q_2^{1,j})$ for $j = 1, \ldots, k_1$ if $l_{1,j} \geq 2$ holds. Finally we have the new paths that leave the cycle at the vertex $e(r_2)$, $(r_1, r_2, q_1^{2,j})$, $j = 1, \ldots, k_2$. So the number of paths p^4 is:

$$a_4 = 1 + k_1 - |\{l_{1,j} \mid l_{1,j} < 2, \; j = 1, \ldots, k_1\}| + k_2.$$

Let $M_{m,i} = |\{l_{m-i,j} \mid l_{m-i,j} < i - 1, \; j = 1, \ldots, k_{m-i}\}|$, then one can extend this to a formula about the number a_m of the paths p^m for all $m \geq 3$:

$$a_m = 1 + \sum_{i=2}^{m-1} k_{m-i} - \sum_{i=3}^{m-1} M_{m,i}.$$

Set $l_{\max} = \max\{l_{ij} \mid \text{for all } i, j\}$ and set x such that $xn + 2 > l_{\max}$. Then

$$a_{xn+3} = 1 + \sum_{i=2}^{xn+2} k_{xn+3-i} - \sum_{i=3}^{xn+2} M_{xn+3,i} = a_{(x+k)n+3}$$

for all $k \geq 1$ because $M_{(x+1)n+3,i} = k_{(x+1)n+3-i}$ for all $xn + 3 \leq i \leq (x+1)n + 3$ by

$$l_{(x+1)n+3-i,j} < i - 1 \quad \text{for all} \quad l_{\max} < xn + 3 \leq i \leq (x+1)n + 3$$

and $r_{(x+1)n+k} = r_{xn+k} = r_k$ for all $k = 1, \ldots, n$.

From $m > xn + 3$ on there are only the paths that got out of the cycle in the last $xn + 2$ steps; all other paths have stopped (except the one along the cycle). So the number of possible paths stays the same after a certain number of passages of the cycles for every further passage and this yields a periodicity. This can be done for every path on the cycle at the beginning because x is independent of p.

If the consideration starts with a relation which is not on a cycle, then there is either a finite path buildable in S or the path gets at last on a cycle and one can continue as above. So there exists either a number N for every $p \in S$ such that $a_{N+k} = 0$ for all $k \geq 0$ or there exist numbers x and n such that $a_{xn+k} = a_{(x+k')n+k}$ for all k, k'. Therefore one finds a number N' that is a period for all $p \in S$ and a general starting level x' of the period.

So the projective modules in the minimal resolution of the algebra are periodic as from a certain level. The differences of the paths that determine the maps in the resolution repeat as one can see from the construction of the paths. They depend only on starting and ending point of the paths and not on the length. Because start and end repeat after x passages of the cycle (or after a multiple in case of more than one cycle) and only the number of passages changes, the maps repeat also periodically subject to m.

Assume now the resolution of A is almost eventually periodic. This means it is infinite. With Lemma 2.2 we know that there is at least one cycle in the quiver of A which is covered by S. If there were two cycles covered and connected by S the resolution would be infinite and not almost periodic (Lemma 2.6). Hence no pair of cycles is connected by S and the resolution is eventually periodic. $\qquad\square$

Corollary 2.8. *Let A be monomial with infinite, but non-periodic resolution. Then there are at least two cycles in the quiver of A which are covered and connected by S.*

Proof. If the resolution is non-periodic, then the quiver must be of one of the following forms by Lemma 2.7:
  (a) There are no cycles in Q. By Lemma 2.2 the resolution is finite.
  (b) There are cycles but no one is covered by S. Thus there are only finite associated sequences so the resolution is here finite, too.
  (c) There are several cycles that are covered and connected by S as in Lemma 2.6. $\qquad\square$

Now we can describe a non-periodic resolution. By construction the number of possible paths $p^n \in AP(n)$ grows by the level n. But there is a certain regularity behind because the paths go along cycles. This means the projective modules will repeat periodically but their multiplicity grows at least linearly. We have already proved the Theorem:

Theorem 2.9. *Let $A = KQ/I$ be a finite dimensional monomial algebra and let I be generated by a minimal set of paths S.*
  (1) A has an infinite, but non-periodic resolution if and only if there are at least two cycles in the quiver of A that are covered and connected by S.
  (2) A has an infinite eventually periodic resolution if and only if
      • there is at least one cycle in Q that is covered by S and
      • no pair of cycles is connected by S.
  (3) A has a finite resolution if and only if there are no cycles in Q that are covered by S.

Corollary 2.10. *Let A be as above. Then the following holds.*

(1) A has a finite resolution if and only if every simple A-module has a finite resolution.

(2) A has an eventually periodic resolution if and only if

- *every simple A-module has an eventually periodic resolution and*
- *at least one simple module has an eventually periodic resolution with a period of positive length.*

(3) A has an infinite, but non-periodic resolution if and only if there exists a simple A-module with an infinite, but non-periodic resolution.

Proof. Follows with Proposition 1.17. □

In particular every simple module that belongs to a vertex in a cycle or to a vertex on a path to a cycle (all covered by S) has an infinite resolution. This resolution is eventually periodic if and only if this cycle is not connected with another one.

Corollary 2.11. *Let A be a monomial algebra. Then A has an eventually periodic resolution if the same holds for A^{op}.*

2.2 Homological ideals

Definition 2.12. Let A and B be two algebras with an algebra epimorphism $\Phi : A \to B$ which means the corresponding functor $\Phi^* : B\text{-Mod} \to A\text{-Mod}$ is full and faithful. Then Φ is an *homological epimorphism* if the corresponding functor on bounded derived categories $D^b(\Phi^*) : D^b(B\text{-Mod}) \to D^b(A\text{-Mod})$ is full and faithful. The kernel of an homological epimorphism is called an *homological ideal*.

In our case there is a nice characterization of homological ideals due to Suarez-Alvarez. He also gave a long exact sequence that relates the Hochschild cohomology of A and A/J which goes back to [10]. We do not need this sequence but state it for completeness. The same sequence and two further ones were independently obtained by König and Nagase [11]. They used stratifying ideals and the bounded derived categories of A and B.

Proposition 2.13. *[13, 5.13] Let Q be a finite quiver with path algebra KQ, let $I < KQ$ be an admissible monomial ideal and put $A = KQ/I$. Let $e \in Q_0$ be a vertex in Q such that no minimal generator of I has e as an inner vertex. Then $J = AeA$ is an homological ideal and*

$$0 \longrightarrow Z(A) \cup J \longrightarrow H^0(A) \to H^0(B) \longrightarrow \operatorname{Ext}^1_A(D(eA), Ae) \longrightarrow \cdots$$
$$\cdots \to \operatorname{Ext}^p_A(D(eA), Ae) \to H^p(A) \to H^p(B) \to \operatorname{Ext}^{p+1}_A(D(eA), Ae) \to \cdots$$

is a long exact sequence relating $H(A)$ and $H(A/J)$ where $A/J = B$ and $Z(A)$ is the center of A.

Moreover, the condition on e is necessary for the ideal to be homological in this case (see [13]). Considering the resolutions of A and $B = A/AeA$ we get the following Theorem.

Theorem 2.14. *Let $A = KQ/I$ be monomial and finite dimensional, let $e \in A$ be a primitive idempotent such that AeA is an homological ideal in A. Then A has an infinite resolution if and only if the same holds for $B = A/AeA$. The resolution of A is eventually periodic if and only if the same holds for B.*

Proof. By Proposition 2.13 the ideal AeA is homological if and only if the vertex of Q associated to e is no inner vertex of any generator of I. In our terminology, if there is no element $p_i \in S_A$ that has e as an inner vertex, where S_A is a minimal set of generators of the ideal I. This means B has almost the same quiver as A but without the vertex e and without all arrows that end or start in e in the quiver of A.

Assume A has an infinite resolution but B has a finite resolution. Then there are no cycles in the quiver of B that are covered by S_B. So e was an inner vertex of a cycle in the quiver of A that was covered by S_A (Lemma 2.2). Thus e must have been an inner vertex of one of the p_i, which is a contradiction.

If B has an infinite resolution, there is a corresponding cycle in the quiver of B. But this cycle is still contained in the quiver of A. By Lemma 2.7 the resolution of A is periodic if there are cycles that are covered by S_A and there is no connection of two cycles by S_A. But everything that S_A covers exists in B. Thus the assertion follows. □

2.3 Periodic algebras

In this section we describe monomial periodic algebras, i.e., with a periodic resolution periodic from level zero. We will show that A is a monomial periodic algebra if and only if A is a direct product of self-injective Nakayama algebras. We compute explicitly the length of the period of the resolution of the algebra (Theorem 2.19) and of the indecomposable modules (Theorem 2.18).

Before we continue let us recall the description of basic connected self-injective Nakayama algebras.

Proposition 2.15. *[3, V.3.8] A basic and connected algebra A that is not isomorphic to K is a self-injective Nakayama algebra if and only if its quiver is of type $\tilde{A}_n$ with linear orientation and $I = KQ^{+l}$ for some $l \geq 2$.*

This means, the quiver of a self-injective Nakayama algebra is an extended Dynkin quiver of type $\tilde{A}_n$, i.e., a single cycle with n vertices. The ideal I is generated by all paths of a certain length l. We assume that A is not semi-simple

We start our considerations with monomial periodic algebras.

Lemma 2.16. *Let $A = KQ/I$ be monomial and periodic and let S be a minimal set of generators of I. Then*

(1) Every vertex of the quiver Q lies on a cycle.

(2) Every vertex is an inner vertex of an element of S.

(3) Every set $AP(n)$ has the same number of elements.

(4) The quiver Q is a product of disjoint cycles.

Proof. (1) Let A have a periodic resolution (P_i, d_i) from the level $p = 0$ with length N, i.e., $P_0 = P_N$. The module P_0 is determined by the set $AP(0) = Q_0$ and the module P_N is determined by the set $AP(N)$. This means

$$P_0 = \bigoplus_{j \in Q_0} P(j, j') \quad \text{and}$$

$$P_N = \bigoplus_{p^N \in AP(N)} P(e(p^N), s(p^N)').$$

So for every vertex $j \in Q_0$ there must be a path p^N with $s(p^N) = e(p^N) = j$. Hence every vertex lies on a cycle.

(2) Let $i \in Q_0$. Because of (1), i is a vertex in a cycle. Let $\alpha : j \to i$ be an arrow in this cycle. Let the path p_j^N be the cycle to the vertex j associated to the sequence $(r_1, \ldots, r_{N-1})$, $e(r_{N-1}) = s(r_1) = j$, as above. So α is the first arrow of the path $r_1 \in S$ and i is an inner vertex of r_1. If $j = i$ and the cycle is a loop, there is no problem because of $l(r_1) \geq 2$.

(3) By (1) and (2), every vertex lies on a cycle and is covered by S. If there is a connection of two cycles by S, the resolution cannot be periodic (Lemma 2.6). So the paths $p^n \in AP(n)$ can only be built inside the separated cycles. Thus there exists for every path $p \in S$ exact one directed path T and exact one way to build an associated sequence of paths. It follows $|AP(2)| = |AP(3)| = |AP(n)|$ for all $n \geq 2$. Because A is periodic, there are isomorphisms $AP(n) \simeq AP(n+N)$ for all $n \geq 0$ and for the length N of the period. So $|AP(0)| = |AP(1)| = |AP(2)|$ holds, too.

(4) By (3), there must be as many arrows as vertices in the quiver. If several cycles are connected, which means they share at least one vertex, there are automatically less vertices then arrows. So Q must be a product of separated cycles without multiple arrows. $\qquad\square$

The following lemma helps us in Theorem 2.18 to calculate a resolution of any indecomposable A-module.

Lemma 2.17. *Let $A = KQ/KQ^{+l}$ where Q is a cycle with n vertices. Then for all $i \in \{1, \ldots, n\}$ and all $k \in \{1, \ldots, l-1\}$ the following holds*

$$\operatorname{rad}^k P(i) \simeq P(i+k)/\operatorname{rad}^{l-k} P(i+k)$$

23

where $P(i) = Ae_i$ is the indecomposable projective module associated to the vertex $i \in Q_0$.

Proof. In this proof we deal with representations of the quiver. So we need only for the proof the following notation: Every representation of the quiver Q with respect to the ideal $I = KQ^{+l}$ is of the form $M = (M(j), M(\alpha), j \in Q_0, \alpha \in Q_1)$ where $M(j)$ is a K-vector space for every $j \in Q_0$ and $M(\alpha)$ is a K-linear map $M(\alpha) : M(j) \to M(j+1)$ for every arrow $\alpha : j \to j+1 \in Q_1$. If $w = \alpha_1 \ldots \alpha_l$ is a path of length l, then $M(w) = M(\alpha_1) \circ \ldots \circ M(\alpha_l) = 0$.

The indecomposable projective modules Ae_i are from now on until the end of the proof denoted by P_i instead of $P(i)$. The vector space of the representation P_i at the vertex j is denoted by $P_i(j) = e_j Ae_i$, the same for any other A-module viewed as a representation.

Let $l = nx + y$ be a decomposition of l with x maximal. Consider all indices modulo n. Without loss of generality we consider only the module P_1 because the quiver is symmetric in any point. Let the arrows in Q be $\alpha_j : j \to j+1 (\mathrm{mod}\, n)$. Then we can write the projective module P_1 associated to the vertex 1 as a representation in the following way:

$$
P_1(j) = \begin{cases} K^{x+1} & j = 1, \ldots, y \\ K^x & j = y+1, \ldots, n \end{cases}
$$

$$
P_1(\alpha_j) = \begin{cases} E_{x+1} & j = 1, \ldots, y-1 \\ E_x & j = y+1, \ldots, n-1 \\ F_x & j = n \\ G_x & j = y \end{cases}
$$

where E_x is the identity matrix on K^x,

$$
F_x = \begin{pmatrix} 0 \ldots 0 \\ E_x \end{pmatrix} : K^x \to K^{x+1}
$$

and

$$
G_x = \begin{pmatrix} 0 \\ E_x & \vdots \\ 0 \end{pmatrix} : K^{x-1} \to K^x.
$$

If $y = 0$, the map $P_1(\alpha_n)$ is the matrix

$$
\begin{pmatrix} 0 \ldots & 0 \\ E_{x-1} & \vdots \\ & 0 \end{pmatrix}.
$$

It is easy to compute that in every of the following modules the maps depend on the above ones. So we omit considering them.

For $k = x'n + y'$ compute $\mathrm{rad}^k P_1$. We differentiate now the cases $y' \leq y$ and $y' > y$. In the first case

$$\mathrm{rad}^k P_1(j) = \begin{cases} K^{x+1-x'-1} & j = 1, \ldots, y' \\ K^{x-x'+1} & j = y'+1, \ldots, y \\ K^{x-x'} & j = y+1, \ldots, n \,. \end{cases}$$

This leads with $l - k = n(x - x') + y - y' = nx'' + y''$ to

$$\mathrm{rad}^{l-k} P_{1+y'}(j) = \begin{cases} K^{x-x''} & j = 1+y', \ldots, y'+y'' \\ K^{x-x''+1} & j = y''+y'+1, \ldots, y+y' \\ K^{x-x''} & j = y+y'+1, \ldots, y' \end{cases}$$

$$= \begin{cases} K^{x'} & j = 1+y', \ldots, y \\ K^{x'+1} & j = y+1, \ldots, y+y' \\ K^{x'} & j = y+y'+1, \ldots, y' \,. \end{cases}$$

Thus the factor module is

$$P_{1+y'}/\mathrm{rad}^{l-k} P_{1+y'}(j) = \begin{cases} K^{x-x'+1} & j = 1+y', \ldots, y \\ K^{x-x'} & j = y+1, \ldots, y' \end{cases} = \mathrm{rad}^k P_1(j).$$

In the second case it looks similarly:

$$\mathrm{rad}^k P_1(j) = \begin{cases} K^{x+1-x'-1} & j = 1, \ldots, y \\ K^{x-x'+1} & j = y+1, \ldots, y' \\ K^{x-x'} & j = y'+1, \ldots, n \,. \end{cases}$$

And with $l - k = n(x - x' - 1) + (n + y - y') = nx'' + y''$ to

$$\mathrm{rad}^{l-k} P_{1+y'}(j) = \begin{cases} K^{x-x''} & j = 1+y', \ldots, y'+y \\ K^{x-x''-1} & j = y+y'+1, \ldots, y''+y' \\ K^{x-x''} & j = y''+y'+1, \ldots, y' \end{cases}$$

$$= \begin{cases} K^{x'+1} & j = 1+y, \ldots, y'+y \\ K^{x'} & j = y+y'+1, \ldots, y \,. \end{cases}$$

Thus the factor module is

$$P_{1+y'}/\mathrm{rad}^{l-k} P_{1+y'}(j) = \begin{cases} K^{x-x'-1} & j = 1+y', \ldots, y \\ K^{x-x'} & j = y+1, \ldots, y' \end{cases} = \mathrm{rad}^k P_1(j).$$

Every further special case like $x = 0$, $y' = 0$ and so on is contained in the above cases, where $K^0 = 0$ and $j = 1, \ldots, y$ for $y = 0$ is empty. $\qquad\square$

Theorem 2.18. *Let $A = KQ/KQ^{+l}$ where Q is the cycle with n vertices and n arrows $\alpha_j : j \to j+1 (\mathrm{mod}\, n)$. Then every indecomposable finite dimensional A-module that is not projective has a periodic projective resolution with a period of length $m = \min\{2\frac{nk}{l} \mid \frac{nk}{l} \in \mathbb{N}\}$.*

Proof. Let $M \in A$-mod be indecomposable and not projective. It is a special property of self-injective Nakayama algebras that every A-module is a factor module of an indecomposable projective module, more explicit ([3, V.3.5]):

Let A be a basic and connected Nakayama algebra and let M be an indecomposable A-module, then there exists an indecomposable projective module $P \in A$-mod and an integer t with $1 \leq t \leq ll(P)$ such that

$$M \simeq P/\mathrm{rad}^t P.$$

Where $ll(M)$ denotes the Loewy length of a module M. This is the least integer $n \geq 0$ such that $\mathrm{rad}^n M = 0$.

We will now show that $M \simeq P(i)/\mathrm{rad}^t P(i)$ has the following minimal projective resolution $(I_j, d_j) \xrightarrow{\pi} M$ with

$$I_j = \begin{cases} P(i + \frac{j}{2} l) & j \text{ even} \\ P(i + \frac{j-1}{2} l + t) & j \text{ odd} \end{cases}$$

and

$$\ker d^j = \begin{cases} \mathrm{rad}^t P(i + \frac{j}{2} l) & j \text{ even} \\ \mathrm{rad}^{l-t} P(i + \frac{j-1}{2} l + t) & j \text{ odd}. \end{cases}$$

By Lemma 2.17 we get $\ker \pi = \mathrm{rad}^t P(i) \simeq P(i+t)/\mathrm{rad}^{l-t} P(i+t)$ and next

$$\ker d^1 = \mathrm{rad}^{l-t} P(i+t) \simeq P(i+l)/\mathrm{rad}^t P(i+l).$$

Assuming the assertion holds for I_j, j even, then it follows that

$$\ker d^j = \mathrm{rad}^t P(i + \frac{j}{2} l) \simeq P(i + \frac{j}{2} l + t)/\mathrm{rad}^{l-t} P(i + \frac{j}{2} l + t).$$

This has the projective cover $P(i + \frac{j}{2} l + t) = I_{j+1}$ with kernel $\mathrm{rad}^{l-t} P(i + \frac{j}{2} l + t) = \ker d^{j+1}$. For j odd we have

$$\ker d^j = \mathrm{rad}^{l-t} P(i + \frac{j-1}{2} l + t) \simeq P(i + \frac{j+1}{2} l)/\mathrm{rad}^t P(i + \frac{j+1}{2} l).$$

This has the projective cover $P(i + \frac{j+1}{2} l) = I_{j+1}$ with kernel $\mathrm{rad}^t P(i + \frac{j+1}{2} l) = \ker d^{j+1}$. By induction the form of the resolution of M is clear.

For $m = 2\frac{nk}{l}$ with $\frac{nk}{l} \in \mathbb{N}$ we have $I_m = P(i + 2\frac{nk}{l} \frac{l}{2}) = P(i+nk) = P(i) = I_0$ because the indices are considered modulo n. If $k = l$, one can find such an $m/2 \in \mathbb{N}$ so the set is not empty and the length of the period is the minimum. Because the length of the period does not depend on t or i, it is the length for every non-projective A-module. $\qquad\square$

Theorem 2.19. *Let A be a finite dimensional monomial algebra.*

(1) Then A is periodic if and only if A is a product of self-injective Nakayama algebras, i.e., $A = \bigoplus_{i=1}^{t} KQ_i/KQ_i^{+l_i}$ with $Q_i = \tilde{A}_{n_i}$ and $l_i \geq 2$.

(2) The length of the period of every subalgebra $A_i = KQ_i/KQ_i^{+l_i}$ is $m_i = \min\{2\frac{n_i k}{l_i} \mid \frac{n_i k}{l_i} \in \mathbb{N}\}$.

(3) The length of the period of A is $m = \operatorname{lcm}\{m_1, \ldots, m_t\}$.

(4) Every indecomposable module in A-mod$_\mathcal{P}$ is periodic with a period of length m_i for some $i \in \{1, \ldots, t\}$.

Proof. (1) Let $A = KQ/I$ be periodic with a period of length N. Then by Lemma 2.16 (4) the quiver Q is a product of disjoint cycles. Moreover by $AP(0) \simeq AP(N)$ we get an element $p_j^N \in AP(N)$ with $s(p_j^N) = e(p_j^N) = j$ for every vertex $j \in Q_0$. Let Q_i be one of the separated cycles of Q. Let the vertices of Q_i be numerated along the arrows such that for any arrow $s(\alpha) + 1 = e(\alpha)$ holds. Let $p_j \in S$ be the generating relation that starts in $j \in (Q_i)_0$. If $l(p_j) > l(p_{j+1})$ is satisfied, the path p_{j+1} is a proper subpath of p_j. But this contradicts the minimality of S. Thus $l(p_{j-1}) \leq l(p_j)$ holds. The same with $l(p_{j-k}) \leq l(p_j)$ for all k. This means for $k = |(Q_i)_0|$, $j - k = j$, too. Therefore all generating relations on the subquiver Q_i must have the same length $l_i = l(p_j)$. Hence $A = \bigoplus_{i=1}^{t} KQ_i/KQ_i^{+l_i}$.

If A is a product of self-injective Nakayama algebras, we can use Theorem 2.9 to get the periodicity. The resolution of A is periodic from the level $p = 0$ because of (4).

(2) Follows by Proposition 1.17 together with Theorem 2.18.

(3) Follows directly from (2).

(4) Every $M \in A$-mod is a direct sum of A_i-modules. So every indecomposable A-module lives on only one of the A_i and can be considered as an A_i-module. With Theorem 2.18 we get the assertion. $\qquad\square$

Chapter 3

Local algebras

In this section we introduce a new tool whose properties and use are central topics of this thesis. To each algebra A we assign a local algebra A_{loc}. We will show that its two-sided resolution reflects properties of the resolution of simple A-modules. We partition the algebras A by *local equivalence* and introduce the terms *locally finite*, *locally periodic*, *locally almost periodic* and *locally infinite* that later on will turn out to be feasible.

3.1 Definitions

Theorem 3.1 (Wedderburn, Malcev). *[12, 11.6] If K is a perfect field and A is a finite dimensional K-algebra, there is a subalgebra B of A such that $A = B \oplus \mathrm{rad}A$. Moreover B is unique up to conjugacy by units of the form $1-w$, where $w \in \mathrm{rad}A$.*

Definition 3.2. Let $A = KQ/I$ be a finite dimensional basic K-algebra with $|Q_0| = n \geq 2$. Then the *local algebra* $A_{loc} = KQ_{loc}/I_{loc}$ is defined by the pullback in the diagram ($A/\mathrm{rad}A \simeq \bigoplus_{i=1}^n K$):

$$
\begin{array}{ccc}
A_{loc} & \overset{g}{\dashrightarrow} & A \\
{\scriptstyle f}\big\downarrow & & \big\downarrow{\scriptstyle \pi} \\
K & \overset{(1,\dots,1)}{\longrightarrow} & A/\mathrm{rad}A
\end{array}
$$

Theorem 3.1 says that $A = \mathrm{rad}A \oplus A/\mathrm{rad}A$, so we can choose $A_{loc} = \mathrm{rad}A \oplus K$. The morphisms are the inclusion $g(a,b) = (a,(b,\dots,b))$ and the projections $\pi(a,b') = b'$ and $f(a,b) = b$. Now the diagram commutes and A_{loc} is a subalgebra of A.

Construct A_{loc} from A

- All vertices of Q are glued together and build the single vertex of Q_{loc}.

- Thus all arrows of $Q_1 = \{\alpha_1, \ldots, \alpha_m\}$ mutate to loops in

$$(Q_1)_{loc} = \{\alpha_1^*, \ldots, \alpha_m^*\}.$$

- All paths $w \in A$ are still contained in A_{loc} but are concatenations of loops $(w^* \in A_{loc})$.
- The ideal I_{loc} consists of
  - all elements $x^* = \sum k_i w_i^*$ with $x = \sum k_i w_i \in I$
  - but in addition of all paths $w^* = w_1^* w_2^*$ with $w_1, w_2 \in A$ that are zero in A by multiplication of the subpaths $(s(w_1) \neq e(w_2))$.

  These additional relations are called *multiplication-relations* for A. In particular there is a multiplication-relation for each arrow $\alpha \in Q_1$ that is no loop: $(\alpha^*)^2 = 0 \in A_{loc}$.

Definition 3.3. Let A and B be finite dimensional algebras. If A_{loc} is isomorphic to B_{loc}, then the algebras are called *locally equivalent*.

Example 3.4. Let A be the following algebra

$$A = K\, 1 \xrightarrow{\alpha_1} 2 \xrightarrow{\alpha_2} 3 \xrightarrow{\alpha_3} 4 \,/\langle \alpha_3 \alpha_2 \alpha_1 \rangle$$

then the associated local algebra is

$$A_{loc} = K \quad 1 \overset{\alpha_1}{\underset{\alpha_3}{\circlearrowright}} \alpha_2 \,/\langle \alpha_3 \alpha_2 \alpha_1, \alpha_1^2, \alpha_2^2, \alpha_3^2, \alpha_1 \alpha_2, \alpha_1 \alpha_3, \alpha_2 \alpha_3, \alpha_3 \alpha_1 \rangle.$$

In this case every relation of length two is a multiplication-relation. We have listed all further locally equivalent algebras in Example 3.24.

The set of multiplication-relations of a local algebra A_{loc} depends on the original algebra A. If we consider A_{loc}, then in general there is more than one locally equivalent algebra (see Example 3.24). So we need to extend the definition. Every element $x \in I_{loc}$ of length 2 is called *multiplication-relation* if there exists a locally equivalent algebra B such that x is a multiplication-relation for B. In the sections 3.3 and 4.1 we will see ways to construct A from A_{loc}. In every step we get a locally equivalent algebra A_i in which a multiplication-relation $x = \alpha\beta$ was split. This means the path $\alpha\beta$ does no longer exist in the quiver of A_i because $s(\alpha) \neq e(\beta)$. We call every multiplication-relation that still exists in A_i a multiplication-relation of A_i.

The modules of the local algebra With the inclusion $g : A_{loc} \to A$ we get the *forgetful functor*

$$-_{loc} : A\text{-mod} \to A_{loc}\text{-mod},$$

by which each A-module can be considered as an A_{loc}-module and each A-module homomorphism can be considered as an A_{loc}-module homomorphism. Let $M \in A\text{-mod}$, then $M_{loc} \in A_{loc}\text{-mod}$ with $a^* \cdot M_{loc} = g(a)M$, thus the underlying vector space of M stays the same for M_{loc}. Let $f : M \to N$ be an A-module homomorphism, then $f_{loc} : M_{loc} \to N_{loc}$ is defined by $f_{loc}(a^* \cdot m) = f(g(a)m) = g(a)f(m) = a^* \cdot f_{loc}(m)$.

Remark 3.5. We should mention that $(A^e)_{loc}$ and $(A_{loc})^e$ are two different algebras. Both algebras have only one simple module. But the quiver of the first one has more arrows than the quiver of the second one; namely all arrows of the forms $\alpha \otimes e'_j$ and $e_i \otimes \alpha'$ for all $\alpha \in Q_1$ and all $i, j \in Q_0$. The algebra $(A_{loc})^e$ has only the arrows $\alpha \otimes e'_1$ and $e_1 \otimes \alpha'$ for all $\alpha \in Q_1$ because A_{loc} has just one vertex e_1. In the following we will only consider $(A_{loc})^e$ and denote it by A^e_{loc}.

3.2 The resolution of the local algebra

In this section we give explicit formulas to compute the resolutions of the A_{loc}-modules in terms of the resolutions of the corresponding A-modules (Theorem 3.10). A special case, which is a little less complicated, is the resolution of the simple A_{loc}-module S_{loc} (Theorem 3.14). This can be used to give a description of the resolution of A_{loc} over A^e_{loc} (Corollary 3.15).

3.2.1 Idea

If one considers the minimal projective resolution of the simple A_{loc}-module S_{loc}, one will notice that the resolutions of the simple A-modules are all contained in that resolution: Let $P_{loc} \in A_{loc}\text{-mod}$ be the single indecomposable projective A_{loc}-module. Then $\mathrm{rad}P_{loc}$ decomposes in

$$\mathrm{rad}P_{loc} = \bigoplus_{i \in Q_0} (\mathrm{rad}P(i))_{loc}$$

because all paths of positive length exist in both algebras, but in A_{loc} they start all in the single vertex of Q_{loc} which we denote by 1. In $\mathrm{rad}P_{loc}$ exactly the idempotent is missing, so the module decomposes in the images of the original radicals under the functor $-_{loc}$. Let

$$0 \to \Omega^2(S(i)) \to \bigoplus_{j \in Q_0} P(j)^{n_j} \to \mathrm{rad}P(i) \to 0$$

be the projective cover of $\operatorname{rad} P(i)$ and its kernel in A, $n_j \geq 0$. And let $\bar{n}_i = \sum n_j$. Then

$$0 \to \Omega^2(S(i))_{loc} \oplus \bigoplus_{j \in Q_0} \left(\bigoplus_{l \neq j} (\operatorname{rad} P(l))_{loc}^{n_j} \right) \to P_{loc}^{\bar{n}_i} \to (\operatorname{rad} P(i))_{loc} \to 0$$

is the projective cover of $(\operatorname{rad} P(i))_{loc}$ in A_{loc}. Because at least one $j \in Q_0$ exists such that $n_j > 0$, at least $x - 1$ summands of the form $(\operatorname{rad} P(l))_{loc}$, $x = |Q_0|$, lie in the kernel of the cover. If $(P_{loc}^{m_k}, d^k)$ is the minimal projective resolution of the simple module $S_{loc} \in A_{loc}$-mod, it follows that $m_0 = 1$, $m_1 = \sum_{i \in Q_0} \bar{n}_i$, $m_2 \geq x - 1$ and

$$m_k \geq (x - 1) m_{k-1} \geq (x - 1)^{k-1}.$$

So the resolution of S_{loc} grows exponentially if $|Q_1| > 1$ and so does the resolution of A_{loc} over A_{loc}^e (Corollary 2.10). If $|Q_1| = 1$, the resolution is periodic.

Every module in A_{loc} can only be covered by P_{loc}^m for some integer m. This yields much bigger kernels of the projective covers than in A-mod. In the next step of the resolution we get much bigger projective modules, namely $P_{loc}^{m'}$. So the original resolution in A-mod exists still in the resolution in A_{loc}-mod but is easy to miss. However this leads to differentiating the term of an infinite, but non-periodic resolution:

Definition 3.6. Let $A = KQ/I$ be a finite dimensional algebra.
(a) • If there exists a finite dimensional connected algebra B that is locally equivalent to A, i.e., $B_{loc} \simeq A_{loc}$, and has finite global dimension, then the structure of the resolution of A over A^e is called *locally finite*.
  • If there is no locally equivalent algebra with a finite resolution but an algebra B with an almost periodic resolution, then the resolution of A is called *locally almost periodic*.
  • If there is only an algebra B with a periodic resolution that is locally equivalent to A but none with finite global dimension, then the resolution of A is called *locally periodic*.
  • If there is neither an algebra that is locally equivalent to A and has finite global dimension nor one with an almost periodic resolution, then the resolution of A is called *locally infinite*.

(b) • If there exists a finite dimensional connected algebra B that is locally equivalent to A and whose simple modules have finite resolutions, then the structure of the resolutions of the simple A-modules are called *locally finite*.
  • If there is no locally equivalent algebra with finite global dimension but an algebra B whose simple modules have almost eventually periodic resolutions, then the resolutions of the simple A-modules are called *locally almost periodic*.

- If there is only a locally equivalent algebra B whose simple modules have eventually periodic resolutions but none with finite global dimension, then the resolutions of the simple A-modules are called *locally periodic*.
- If there is neither an algebra that is locally equivalent to A and has finite global dimension nor one whose simple modules have almost periodic resolutions, then the resolutions of the simple A-modules are called *locally infinite*.

Remark 3.7. In the monomial case the terms locally periodic and locally almost periodic are equivalent (compare Lemma 2.7). In the non-monomial case there are only some inequalities (see Proposition 4.13). By definition, the resolution of A is locally finite if and only if the resolutions of the simple A-modules are locally finite.

3.2.2 The resolution of the A_{loc}-modules

In the following we give an explicit calculation of the minimal resolution of a module $M_{loc} \in A_{loc}$-mod in relation to the resolution of $M \in A$-mod. We need to fix some notation.

Notation. Let $A = KQ/I$ be a finite dimensional algebra and let $M \in A$-mod be a module with minimal resolution (P_i, d_i), the projectives are $P_i = \bigoplus_{j \in Q_0} P(j)^{n_j^i}$ with $n_j^i \geq 0$. Denote the minimal projective resolutions of the simple A-modules $S(j)$ by (P_i^j, d_i^j) with $P_i^j = \bigoplus_{k \in Q_0} P(k)^{j m_i^k}$ and ${}^j m_i^k \geq 0$. Moreover

$$n_i := \sum_{j \in Q_0} n_j^i \quad \text{and} \quad m_i^j := \sum_{k \in Q_0} {}^j m_i^k.$$

Now we get some projective covers in A_{loc}-mod:

The cover of M_{loc}:

$$0 \to \Omega^1(M)_{loc} \oplus \bigoplus_{j \in Q_0} \bigoplus_{i \neq j} \mathrm{rad}P(i)_{loc}^{n_j^0} \to P_{loc}^{n_0} \to M_{loc} \to 0 \tag{3.1}$$

Where the kernel consists of the image of the kernel of the cover of M together with the remaining part of the radical of $P_{loc}^{n_0}$. The following is an easy computation

$$M_0 := \bigoplus_{j \in Q_0} \bigoplus_{i \neq j} \mathrm{rad}P(i)_{loc}^{n_j^0} = \bigoplus_{i \in Q_0} \mathrm{rad}P(i)_{loc}^{(n_0 - n_i^0)} = \bigoplus_{i \in Q_0} \Omega^1(S(i))_{loc}^{(n_0 - n_i^0)}.$$

The cover of $\Omega^k(M)_{loc}$: More generally we get

$$0 \to \Omega^{k+1}(M)_{loc} \oplus M_k \to P_{loc}^{n_k} \to \Omega^k(M)_{loc} \to 0$$

with $M_k := \bigoplus_{j \in Q_0} \bigoplus_{i \neq j} \mathrm{rad}P(i)_{loc}^{n_j^k} = \bigoplus_{i \in Q_0} \Omega^1(S(i))_{loc}^{(n_k - n_i^k)}.$

The cover of $\Omega^k(S(l))_{loc}$:

$$0 \to \Omega^{k+1}(S(l))_{loc} \oplus \bigoplus_{j \in Q_0} \bigoplus_{i \neq j} \operatorname{rad} P(i)_{loc}^{\,^l m_k^j} \to P_{loc}^{m_k^l} \to \Omega^k(S(l))_{loc} \to 0.$$

Set

$$N(k,l) := \Omega^k(S(l))_{loc} \oplus \bigoplus_{i \in Q_0} \Omega^1(S(i))_{loc}^{(m_{k-1}^l - \,^l m_{k-1}^i)} = \Omega^1(\Omega^k(S(l))_{loc}).$$

The cover of M_k: With $m_k := \sum_{i \in Q_0}(n_k - n_i^k)m_1^i$ the module M_k is covered by $P_{loc}^{m_k}$ with kernel

$$\bigoplus_{i \in Q_0} \left[\Omega^2(S(i))_{loc} \oplus \bigoplus_{l \in Q_0} \operatorname{rad} P(l)_{loc}^{(m_1^i - \,^i m_1^l)} \right]^{(n_k - n_i^k)}$$

$$= \bigoplus_{i \in Q_0} N(2,i)^{(n_k - n_i^k)}.$$

The cover of $N(j,i)$: With $n(j,i) := m_j^i + \sum_{l \in Q_0}(m_{j-1}^i - \,^i m_{j-1}^l)m_1^l$ the module $N(j,i)$ is covered by $P_{loc}^{n(j,i)}$ with kernel

$$N(j+1,i) \oplus \bigoplus_{l \in Q_0} N(2,l)^{(m_{j-1}^i - \,^i m_{j-1}^l)}.$$

The modules $O(x,y,v)$: Define recursively

$$o_{k_{t+1}}(y,(x_1,\dots,x_t)) := \sum_{k_t \in Q_0} (m_{x_t}^{k_t} - \,^{k_t} m_{x_t}^{k_{t+1}}) o_{k_t}(y,(x_1,\dots,x_{t-1}))$$

with $o_{k_1}(y,()) := n_y - n_{k_1}^y$ and set

$$O(x,y,(x_1,\dots,x_t)) := \bigoplus_{k_{t+1} \in Q_0} N(x,k_{t+1})^{o_{k_{t+1}}(y,(x_1,\dots,x_t))}.$$

We get $O(x,y,(x_1)) = \bigoplus_{k_2 \in Q_0} N(x,k_2)^{o_{k_2}(y,(x_1))}$ and $\Omega^1(M_k) = O(2,k,())$.

The cover of $O(x,y,v)$: With

$$o(x,y,(x_1,\dots,x_t)) := \sum_{k_{t+1} \in Q_0} n(x,k_{t+1}) o_{k_{t+1}}(y,(x_1,\dots,x_t))$$

the module $O(x,y,(x_1,\dots,x_t))$ is covered by $P_{loc}^{o(x,y,(x_1,\dots,x_t))}$ with kernel

$$\bigoplus_{k_{t+1} \in Q_0} \left[N(x+1,k_{t+1}) \oplus \bigoplus_{k \in Q_0} N(2,k)^{(m_{x-1}^{k_{t+1}} - \,^{k_{t+1}} m_{x-1}^k)} \right]^{o_{k_{t+1}}(y,(x_1,\dots,x_t))}$$

$$= O(x+1,y,(x_1,\dots,x_t)) \oplus O(2,y,(x_1,\dots,x_t,x-1)).$$

The case $O(x, y, (\,))$: Now we see that the definition of $O(x, y, v)$ makes sense even in the case $v = (\,)$: $o(x, y, (\,)) = \sum_{k_1 \in Q_0} n(x, k_1)^{(n_y - n_{k_1}^y)}$ and the kernel of the cover of $\bigoplus_{i \in Q_0} N(x, i)^{(n_y - n_i^y)}$ is

$$\bigoplus_{i \in Q_0} N(x+1, i)^{(n_y - n_i^y)} \oplus O(2, y, (x-1))$$
$$= O(x+1, y, (\,)) \oplus O(2, y, (x-1))$$
$$= \Omega^1(O(x, y, (\,))).$$

The decomposition of $\Omega^i(M_{loc})$: One can see that every $\Omega^i(M_{loc})$ can be decomposed in modules of the above forms. In the following diagram the i-th row denotes the decomposition of the module $\Omega^i(M_{loc})$ where $O(x, y, (\,))$ is abbreviated by $O(x, y)$. There are five ways of building the summands, each encoded by 1 to 5:

1$=$ $\Omega^k(M)_{loc}$ as part of the kernel of $\Omega^{k-1}(M)_{loc}$.
2$=$ M_{k-1} as part of the kernel of $\Omega^{k-1}(M)_{loc}$.
3$=$ $O(2, k-1, (\,))$ as the kernel of M_{k-1}.
4$=$ $O(k, y, (x_1, \ldots, x_t))$ as part of the kernel of $O(k-1, y, (x_1, \ldots, x_t))$.
5$=$ $O(2, y, (x_1, \ldots, x_t, k-2))$ as part of the kernel of $O(k-1, y, (x_1, \ldots, x_t))$.

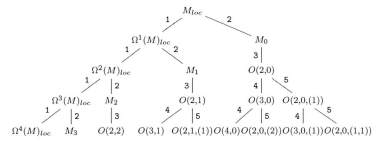

And so on. Every summand of $\Omega^i(M_{loc})$ has a unique sequence of length i of the steps $1, 2, 3, 4, 5$ by that it was constructed. For example $\Omega^i(M)_{loc}$ belongs to the sequence $(1, \ldots, 1)$. The module M_{i-1} belongs to the sequence $(1, \ldots, 1, 2)$ and M_0 belongs to (2). Let 1^k, 4^k, and 5^k denote a series of lengths k of 1, 4 or 5, respectively. Then all sequences of the form

$$x_p = (1^{k_1}, 2, 3, 4^{k_1^4}, 5^{k_1^5}, 4^{k_2^4}, \ldots, 4^{k_p^4}, 5^{k_p^5})$$

with $k_1 \geq 0$, $k_i^j \geq 0$ and $k_1 + 2 + \sum_j (k_j^4 + k_j^5) = i$ belong to a summand of the module $\Omega^i(M_{loc})$.

For notation say if a sequence is called x_p with $p \geq 0$, then $k_p^4 > 0$ must hold. Let $\bar{S}(i)$ be the set of all sequences x_p of the above form of length $i \geq 2$. Let $\bar{S}_4(i)$ be the set of all sequences $x_p \in \bar{S}(i)$ with $k_p^5 = 0$ and let $\bar{S}_5(i)$ be the set of all sequences with $k_p^5 > 0$. Thus $x_0 = (1^{i-2}, 2, 3) \in \bar{S}_4(i)$ and belongs to $O(2, i-2, (\,))$ for all $i \geq 2$.

Let $T(i)$ denote the subset of $\bar{S}(i)$ containing only x_p with $p > 0$, the same for $T_4(i)$ and $T_5(i)$.

Lemma 3.8. *Let* $x_p = (1^{k_1}, 2, 3, 4^{k_1^4}, 5^{k_1^5}, 4^{k_2^4}, \ldots, 4^{k_p^4}, 5^{k_p^5}) \in \bar{S}(i)$ *with* $p > 0$, *then the associated summand of* $\Omega^i(M_{loc})$ *is*

- $O(k_p^4 + 2, k_1, v_{p-1})$ *if* $x_p \in \bar{S}_4(i)$ *or*
- $O(2, k_1, v_p)$ *if* $x_p \in \bar{S}_5(i)$.

Where the vector is $v_0 = (\,)$ *or the vector of length* $\sum_{l=1}^j k_l^5$ *for* $0 < j \in \{p-1, p\}$

$$v_j = (k_1^4 + 2, 1^{k_1^5 - 1}, k_2^4 + 2, 1^{k_2^5 - 1}, \ldots, k_j^4 + 2, 1^{k_j^5 - 1}).$$

Proof. Induction on i. Supposing $p > 0$ means $k_p^4 > 0$, so we start the induction at $i = 3$. Hence $T(3) = \{x_1 = (2, 3, 4)\}$ and the associated summand is as one can see in the diagram above $O(3, 0, (\,)) = O(k_1^4 + 2, k_1, v_0)$.

For a better understanding, we consider first the case $i = 4$ before doing the induction step. Thus $T_4(4) = \{x_1^1 = (1, 2, 3, 4), x_1^2 = (2, 3, 4, 4), x_2 = (2, 3, 5, 4)\}$ where x_1^1 belongs to $O(3, 1, (\,))$, x_1^2 belongs to $O(4, 0, (\,))$ and x_2 belongs to $O(3, 0, (1))$ (compare the diagram). The other sequences are

$$T_5(4) = \{y_1^1 = (1, 2, 3, 5), y_1^2 = (2, 3, 5, 5), y_1^3 = (2, 3, 4, 5)\};$$

$\to y_1^1$ belongs to $O(2, 1, (1)) = O(2, k_1, (k_1^4 + 2))$
$\to y_1^2$ belongs to $O(2, 0, (1, 1)) = O(2, k_1, (k_1^4 + 2, 1^{k_1^5 - 1}))$
$\to y_1^3$ belongs to $O(2, 0, (2)) = O(2, k_1, (k_1^4 + 2))$

Assume the assertion is true for i and consider $T(i+1)$. Then every $x_p \in T(i+1)$ can be written as either $(\bar{x}_p, 4)$ or $(\bar{x}_p, 5)$ for some $\bar{x}_p \in T(i)$. Let $x_p \in T_4(i+1)$ then $x_p = (\bar{x}_p, 4)$. If $\bar{x}_p \in T_4(i)$, then the associated summand of $\Omega^i(M_{loc})$ is $O(\bar{k}_p^4 + 2, \bar{k}_1, \bar{v}_{p-1})$ with

$$\bar{v}_{p-1} = (\bar{k}_1^4 + 2, 1^{\bar{k}_1^5 - 1}, \bar{k}_2^4 + 2, 1^{\bar{k}_2^5 - 1}, \ldots, \bar{k}_{p-1}^4 + 2, 1^{\bar{k}_{p-1}^5 - 1}).$$

Hence $k_p^4 = \bar{k}_p^4 + 1$ and we get the summand $O(\bar{k}_p^4 + 2 + 1, \bar{k}_1, \bar{v}_{p-1}) = O(k_p^4 + 2, k_1, v_{p-1})$.

If $\bar{x}_p \in T_5(i)$, we should write x_{p+1} because $k_{p+1}^4 = 1$ and we get the associated summand $O(2, \bar{k}_1, \bar{v}_p)$ of $\Omega^i(M_{loc})$. The next construction step 4 yields $O(3 = k_{p+1}^4 + 2, \bar{k}_1, \bar{v}_p)$ as expected.

Let now $x_p \in T_5(i+1)$ with $x_p = (\bar{x}_p, 5)$. If $\bar{x}_p \in T_4(i)$, we have the associated summand $O(\bar{k}_p^4 + 2, \bar{k}_1, \bar{v}_{p-1})$ of $\Omega^i(M_{loc})$. The next construction step is 5 so we get $O(2, \bar{k}_1, (\bar{v}_{p-1}, \bar{k}_p^4 + 2)) = O(2, k_1, v_p)$ with $k_p^5 = 1$.

If $\bar{x}_p \in T_5(i)$, we have the associated summand $O(2, \bar{k}_1, \bar{v}_p)$ of $\Omega^i(M_{loc})$ with $\bar{v}_p = (\bar{k}_1^4 + 2, 1^{\bar{k}_1^5 - 1}, \bar{k}_2^4 + 2, 1^{\bar{k}_2^5 - 1}, \ldots, \bar{k}_p^4 + 2, 1^{\bar{k}_p^5 - 1})$. In the next construction step we get $O(2, \bar{k}_1, (\bar{v}_p, 1))$ with $k_p^5 = \bar{k}_p^5 + 1$. Thus $v_p = (\bar{k}_1^4 + 2, 1^{\bar{k}_1^5 - 1}, \bar{k}_2^4 + 2, 1^{\bar{k}_2^5 - 1}, \ldots, \bar{k}_p^4 + 2, 1^{k_p^5 - 1})$. $\square$

Now we can give a concrete description of $\Omega^i(M_{loc})$.

Lemma 3.9.

$$\Omega^i(M_{loc}) = \Omega^i(M)_{loc} \oplus M_{i-1}$$
$$\oplus \bigoplus_{x_p \in T_4(i)} O(k_p^4 + 2, k_1, v_{p-1})$$
$$\oplus \bigoplus_{x_p \in T_5(i)} O(2, k_1, v_p).$$

for all $i \geq 1$ with v_p as in Lemma 3.8.

Proof. Induction on i. For $i = 1$ see (3.1). Assume the assertion holds for $\Omega^i(M_{loc})$, then we have

$$\Omega^{i+1}(M_{loc}) = \Omega^1(\Omega^i(M_{loc}))$$
$$= \Omega^{i+1}(M)_{loc} \oplus M_i \oplus O(2, i-2, ())$$
$$\oplus \bigoplus_{x_p \in T_4(i)} [O(k_p^4 + 3, k_1, v_{p-1}) \oplus O(2, k_1, (v_{p-1}, k_p^4 + 2))]$$
$$\oplus \bigoplus_{x_p \in T_5(i)} [O(3, k_1, v_p) \oplus O(2, k_1, (v_p, 1))]$$

As in Lemma 3.8 and its proof we get for each summand a sequence $x_p \in \bar{S}(i+1)$ (and for each sequence in $\bar{S}(i + 1)$ a summand):

$$
\begin{array}{rl}
O(2, i-2, ()) & \text{belongs to} \quad (1^{i-1}, 2, 3) \\
O(k_p^4 + 3, k_1, v_{p-1}) & \text{belongs to} \quad (x_p, 4) \text{ with } x_p \in \bar{S}_4(i) \\
O(2, k_1, (v_{p-1}, k_p^4 + 2)) & \text{belongs to} \quad (x_p, 5) \text{ with } x_p \in \bar{S}_4(i) \\
O(3, k_1, v_p) & \text{belongs to} \quad (x_p, 4) \text{ with } x_p \in \bar{S}_5(i) \\
O(2, k_1, (v_p, 1)) & \text{belongs to} \quad (x_p, 5) \text{ with } x_p \in \bar{S}_5(i)
\end{array}
$$

By reordering the above sums into $T_4(i+1)$ and $T_5(i+1)$ we get the assertion. $\quad\square$

Theorem 3.10. *Let A be finite dimensional and let $M \in A$-mod. Then with the above notation the minimal projective resolution of the module $M_{loc} \in A_{loc}$-mod is of the form $P_i = P_{loc}^{x_i}$ with*

$$x_i = n_i + m_{i-1} + \sum_{x_p \in T_4(i)} o(k_p^4 + 2, k_1, v_{p-1}) + \sum_{x_p \in T_5(i)} o(2, k_1, v_p).$$

The maps $d_{i+1} : P_{i+1} \to P_i$ are the projective covers of the modules $\Omega^i(M_{loc})$, which can be written as

$$\Omega^i(M_{loc}) = \Omega^i(M)_{loc} \oplus \bigoplus_{l \in Q_0} \bigoplus_{j=1}^{i} \Omega^j(S(l))_{loc}^{s_l^j}.$$

with

$$s_l^1 = o_l(i-1,()) + \sum_{x_p \in T_5(i+1)} o_l(k_1, v_p)$$

$$s_l^2 = o_l(i-1,()) + \sum_{x_p \in T_5(i)} o_l(k_1, v_p)$$

and for all $3 \leq k \leq i$

$$s_l^k = \sum_{x_{p-1} \in T_5(i-k+2)} o_l(k_1, v_{p-1}).$$

Where for $x_p \in T_4(i)$ the length of v_{p-1} is $t = \sum_{j=1}^{p-1} k_j^5$ and for $x_p \in T_5(i)$ it is $t = \sum_{j=1}^{p} k_j^5$.

Remark 3.11. With this Theorem we have a minimal projective resolution for every A_{loc}-module M_{loc}, where M_{loc} is the image of an A-module M under the forgetful functor $-_{loc}$ (compare 3.1). This functor does not change the underlying vector space of the module but the module structure with

$$a^* M_{loc} = g(a)M$$

for $a^* \in A_{loc}$ and the inclusion $g : A_{loc} \to A$. Thus if we consider A as an A^e-module, then the $(A^e)_{loc}$-module A_{loc} is not equal to the local algebra A_{loc}. We already know $(A^e)_{loc} \not\cong (A_{loc})^e$.

This means the projective resolution of $A_{loc} \in (A^e)_{loc}$-mod we get by Theorem 3.10 has nothing to do with the projective resolution of the local algebra A_{loc} considered as an $(A_{loc})^e$-module. Only the last resolution can be used to compute the Hochschild cohomology of A_{loc}. But we can use the resolution of the simple module S_{loc} of A_{loc} and Lemma 1.3. See Corollary 3.15.

Proof. The exponents of the projective modules in the resolutions follow directly from Lemma 3.18. The decomposition of $\Omega^i(M_{loc})$ is as follows. We start with

$$\Omega^i(M_{loc}) = \Omega^i(M)_{loc} \oplus M_{i-1}$$
$$\oplus \bigoplus_{x_p \in T_4(i)} O(k_p^4 + 2, k_1, v_{p-1})$$
$$\oplus \bigoplus_{x_p \in T_5(i)} O(2, k_1, v_p).$$

By definition $M_{i-1} = \bigoplus_{l \in Q_0} \Omega^1(S(l))_{loc}^{o_l(i-1,())}$, which gives us a first summand of the s_l^1's. Next consider the summands $O(k_p^4 + 2, k_1, v_{p-1})$ of $\Omega^i(M_{loc})$ associated to $x_p \in T_4(i)$

$$O(k_p^4 + 2, k_1, v_{p-1})$$

$$= \bigoplus_{k_{t+1} \in Q_0} N(k_p^4 + 2, k_{t+1})^{o_{k_{t+1}}(k_1, v_{p-1})}$$

$$= \bigoplus_{k_{t+1} \in Q_0} \left[\Omega^{k_p^4 + 2}(S(k_{t+1}))_{loc} \oplus \bigoplus_{l \in Q_0} \Omega^1(S(l))_{loc}^{(m_{k_p^4+1}^{k_{t+1}} - {}^{k_{t+1}}m_{k_p^4+1}^l)} \right]^{o_{k_{t+1}}(k_1, v_{p-1})}$$

From this we get a second summand of s_l^1. Considering the summands of $\Omega^i(M_{loc})$ associated to $x_p \in T_5(i)$ we get

$$O(2, k_1, v_p)$$

$$= \bigoplus_{k_{t+1} \in Q_0} N(2, k_{t+1})^{o_{k_{t+1}}(k_1, v_p)}$$

$$= \bigoplus_{k_{t+1} \in Q_0} \left[\Omega^2(S(k_{t+1}))_{loc} \oplus \bigoplus_{l \in Q_0} \Omega^1(S(l))_{loc}^{(m_1^{k_{t+1}} - {}^{k_{t+1}}m_1^l)} \right]^{o_{k_{t+1}}(k_1, v_p)}$$

This yields a third summand of s_l^1 and we have considered all summand of $\Omega^i(M_{loc})$, which means we have collected all $\Omega^1(S(l))_{loc}$'s. We get

$$
\begin{aligned}
s_l^1 &= o_l(i-1, ()) \\
&+ \sum_{x_p \in T_4(i)} \sum_{k_{t+1} \in Q_0} (m_{k_p^4+1}^{k_{t+1}} - {}^{k_{t+1}}m_{k_p^4+1}^l) o_{k_{t+1}}(k_1, v_{p-1}) \\
&+ \sum_{x_p \in 5(i)} \sum_{k_{t+1} \in Q_0} (m_1^{k_{t+1}} - {}^{k_{t+1}}m_1^l) o_{k_{t+1}}(k_1, v_p) \\
&= o_l(i-1, ()) \\
&+ \sum_{x_p \in 4(i)} o_l(k_1, (v_{p-1}, k_p^4 + 1)) + \sum_{x_p \in 5(i)} o_l(k_1, (v_p, 1)) \\
&= o_l(i-1, ()) + \sum_{x_p \in \bar{S}_5(i+1)} o_l(k_1, v_p).
\end{aligned}
$$

The modules $\Omega^2(S(l))_{loc}$ occur in $O(k_p^4 + 2, k_1, v_{p-1})$ if and only if $k_p^4 = 0$, which only holds for $x_0 = (1^{i-2}, 2, 3)$ belonging to $O(2, i-2, ())$. This yields the first summand of s_l^2, the second comes from $O(2, k_1, v_p)$.

We need to collect all $x_p \in T_4(i)$ with the same k_p^4 to get $s_l^{k_p^4+2}$. Every such x_p can be written as $(x_{p-1}, 4^{k_p^4})$ with $x_{p-1} \in T_5(i - k_p^4)$ belonging to $O(2, k_1, v_{p-1})$. Then the k_1's and k_j^5's are the same just as v_{p-1}. Hence the assertion follows. $\square$

3.2.3 The resolution of the simple A_{loc}-module

Special cases of Theorem 3.10 are the simple modules of A, which all map to the single simple module S_{loc} of A_{loc}. Thus the resolution of S_{loc} is independent of the simple A-module we started with. But if we use Theorem 3.10 to compute the resolution S_{loc}, we get a different decomposition for every $S(i) \in A$-mod. In the following we simplify every step of the calculation above for the case S_{loc}.

Notation. Again, we denote the minimal projective resolutions of the simple A-modules $S(j)$ by (P_i^j, d_i^j) with $P_i^j = \bigoplus_{k \in Q_0} P(k)^{j m_i^k}$ and ${}^j m_i^k \geq 0$. Moreover

$$m_i^j := \sum_{k \in Q_0} {}^j m_i^k \quad \text{and} \quad m_i := \sum_{j \in Q_0} m_i^j.$$

The cover of S_{loc}: With $\bigoplus_{i \in Q_0} \Omega^1(S(i))_{loc} = (\mathrm{rad}\, A)_{loc} = \mathrm{rad}\, A_{loc} = \Omega^1(S_{loc})$ we get the projective cover of $S_{loc} \in A_{loc}$-mod:

$$0 \to \bigoplus_{i \in Q_0} \Omega^1(S(i))_{loc} \to P_{loc} \to S_{loc} \to 0.$$

The cover of $\Omega^k(S(i))_{loc}$: In A-mod the modules $\Omega^k(S(i))$ are covered by m_k^i projective modules. There is only one projective A_{loc}-module, we get

$$0 \to \Omega^{k+1}(S(i))_{loc} \oplus \bigoplus_{j \in Q_0} \bigoplus_{l \neq j} \Omega^1(S(l))_{loc}^{{}^i m_k^j} \to P_{loc}^{m_k^i} \to \Omega^k(S(i))_{loc} \to 0.$$

The kernel consists of the image of the kernel of the cover of $\Omega^k(S(i))$ together with the remaining part of the radical of P_{loc}. The following is an easy computation:

$$\bigoplus_{j \in Q_0} \bigoplus_{l \neq j} \Omega^1(S(l))_{loc}^{{}^i m_k^j} = \bigoplus_{j \in Q_0} \Omega^1(S(j))_{loc}^{m_k^i - {}^i m_k^j}.$$

More modules and covers: Again, we have to give some easier names to the occurring modules to be able to formulate Theorem 3.14. Every such module in the following will be given together with its projective cover and the associated kernel.

- $M_1 := \bigoplus_{i \in Q_0} \Omega^1(S(i))_{loc} = \Omega^1(S_{loc})$ is covered by $P_{loc}^{m_1}$ with kernel

$$\bigoplus_{i \in Q_0} \left[\Omega^2(S(i))_{loc} \oplus \bigoplus_{j \in Q_0} \Omega^1(S(j))_{loc}^{m_1^i - {}^i m_1^j} \right] = \bigoplus_{i \in Q_0} N(2, i) = O(2, ())$$

- $N(k,i) := \Omega^k(S(i))_{loc} \oplus \bigoplus_{j \in Q_0} \Omega^1(S(j))_{loc}^{(m_{k-1}^i - {}^i m_{k-1}^j)}$ is covered by $P_{loc}^{n(k,i)}$
  with

$$n(k,i) := m_k^i + \sum_{j \in Q_0} m_1^j (m_{k-1}^i - {}^i m_{k-1}^j)$$

and with kernel

$$N(k+1,i) \oplus \bigoplus_{j \in Q_0} N(2,j)^{(m_{k-1}^i - {}^i m_{k-1}^j)}.$$

The modules $O(x,v)$: Finally, we want to define the modules $O(x,v)$, the summands of $\Omega^i(S_{loc})$ analogue to the modules $O(x,y,v)$ in the general case. Set recursively for $x_i \in \mathbb{N}$

$$o_{k_{t+1}}(x_1,\ldots,x_t) := \sum_{k_t \in Q_0} (m_{x_t}^{k_t} - {}^{k_t} m_{x_t}^{k_{t+1}}) o_{k_t}(x_1,\ldots,x_{t-1})$$

with $o_i(\ ) := 1$. Moreover let

$$o(x,(x_1,\ldots,x_t)) := \sum_{k_{t+1} \in Q_0} o_{k_{t+1}}(x_1,\ldots,x_t) n(x,k_{t+1})$$

and define

$$O(x,(x_1,\ldots,x_t)) := \bigoplus_{k_{t+1} \in Q_0} N(x,k_{t+1})^{o_{k_{t+1}}(x_1,\ldots,x_t)}$$

and $O(x,(\ )) := \bigoplus_{i \in Q_0} N(x,i)$ for $t = 0$.

The cover of $O(x,v)$: The modules $O(x,v)$ are covered by $P_{loc}^{o(x,v)}$ with $v = (x_1,\ldots,x_t)$ and kernel

$$\bigoplus_{k_{t+1} \in Q_0} \left[N(x+1,k_{t+1}) \oplus \bigoplus_{k_{t+2} \in Q_0} N(2,k_{t+2})^{(m_{x-1}^{k_{t+1}} - {}^{k_{t+1}} m_{x-1}^{k_{t+2}})} \right]^{o_{k_{t+1}}(x_1,\ldots,x_t)}$$
$$= O(x+1,(x_1,\ldots,x_t)) \oplus O(2,(x_1,\ldots,x_t,x-1))$$

The case $O(x,(\ ))$: Compare $O(x,(\ )) = \bigoplus_{i \in Q_0} N(x,i)$. We get $o(x,(\ )) = \sum_{i \in Q_0} n(x,i)$ and $P_{loc}^{o(x,(\ ))}$ is the projective cover with kernel

$$\bigoplus_{i \in Q_0} \left[N(x+1,i) \oplus \bigoplus_{j \in Q_0} N(2,j)^{(m_{x-1}^i - {}^i m_{x-1}^j)} \right]$$
$$= O(x+1,(\ )) \oplus O(2,(x-1))$$

41

The decomposition of $\Omega^i(S_{loc})$: Now we can see that every $\Omega^i(S_{loc})$ can be decomposed in modules of the above forms; the summands can be encoded by the way they were built as kernels. Most of the summands are of type $O(x, (x_1, \ldots, x_t))$, the kernel of the projective cover of such modules can always be written as a sum of two modules of the same type. Denote the way to build the summand $O(x+1, (x_1, \ldots, x_t))$ of the kernel by 2 and the other summand $O(2, (x_1, \ldots, x_t, x-1))$ by 3. In the following diagram the i-th row denotes the decomposition of the module $\Omega^i(S_{loc})$.

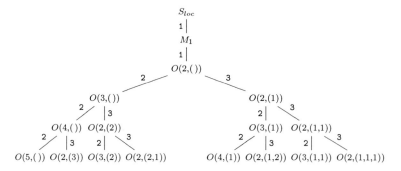

And so on. Every summand of $\Omega^i(S_{loc})$ has a unique sequence of length i of the steps 1,2,3 by that it was constructed. For example $O(x, ())$ belongs to the sequence $(1, 1, 2^{x-2})$ and is a summand of $\Omega^x(S_{loc})$. Here $2^k, 3^k$ denote sequences of length k of 2 or 3. Every sequence is of the form

$$x_p = (1, 1, 2^{k_1^2}, 3^{k_1^3}, 2^{k_2^2}, \ldots, 2^{k_p^2}, 3^{k_p^3}).$$

Let $S(i)$ be the set of all sequences x_p of length i of the above form, $i \geq 1$. Let $S_2(i) \subset S(i)$ be the set of all sequences x_p with $k_p^3 = 0$, thus $S_2(2) = \{x_0 = (1, 1)\}$. Let $S_3(i) \subset S(i)$ be the set of all sequences x_p with $k_p^3 > 0$, thus $S_3(2) = \emptyset$ and $S_3(3) = \{(1, 1, 3)\}$. We have $S(1) = \{(1)\}$, $S(2) = \{(1, 1)\}$ and $S(3) = \{(1, 1, 2), (1, 1, 3)\}$.

Lemma 3.12. *Let* $x_p = (1, 1, 2^{k_1^2}, 3^{k_1^3}, 2^{k_2^2}, \ldots, 2^{k_p^2}, 3^{k_p^3}) \in S(i)$ *with* $p \geq 0$, *then the associated summand of* $\Omega^i(S_{loc})$ *is either* $O(2, ())$ *if* $p = 0$ *or for* $p > 0$ *either*

- $O(k_p^2 + 2, v_{p-1})$ *for* $x_p \in S_2(i)$ *or*
- $O(2, v_p)$ *for* $x_p \in S_3(i)$.

The vector is $v_j = (k_1^2 + 1, 1^{k_1^3 - 1}, k_2^2 + 1, \ldots, k_j^2 + 1, 1^{k_j^3 - 1})$ *of length* $i = \sum_{l=1}^{j} k_l^3$ *for* $j \in \{p - 1, p\}$.

Proof. Induction on i. The case $i = 2$ and $p = 0$ is clear (compare the diagram above). For $i = 3$ there is only one sequence $x_1 = (1, 1, 3)$ in the set $S_3(3)$ and it belongs to $O(2, (1))$. The set $S_2(3)$ contains also just one sequence, namely $(1, 1, 2)$, which belongs to $O(3, ())$.

Assume, the assertion is true for i and consider $S(i+1)$. Then every $x_p \in S(i+1)$ can be written as either $(\bar{x}_p, 2)$ or $(\bar{x}_p, 3)$ for some $\bar{x}_p \in S(i)$. If $k_p^2 = 1$ for x_p, then we should write $x_p = (\bar{x}_{p-1}, 2)$ and $\bar{x}_{p-1} \in S_3(i)$. In every case the proof is the same as in Lemma 3.8 just change 2 into 4 and 3 into 5. $\qquad\square$

Lemma 3.13. *For all $i \geq 2$ and with v_p as in Lemma 3.12:*

$$\Omega^i(S_{loc}) = \bigoplus_{x_p \in S_2(i)} O(k_p^2 + 2, v_{p-1}) \oplus \bigoplus_{x_p \in S_3(i)} O(2, v_p).$$

Proof. Induction on i. In case $i = 2$ compare the diagram. Assume the assertion holds for i, we get

$$
\begin{aligned}
\Omega^{i+1}(S_{loc}) &= \Omega^1(\Omega^i(S_{loc})) \\
&= \bigoplus_{x_p \in S_2(i)} \left[O(k_p^2 + 3, v_{p-1}) \oplus O(2, (v_{p-1}, k_p^2 + 1)) \right] \oplus \\
&\qquad \bigoplus_{x_p \in S_3(i)} \left[O(3, v_p) \oplus O(2, (v_p, 1)) \right].
\end{aligned}
$$

By reordering the above sums into $S_2(i+1)$ and $S_3(i+1)$ we get the assertion. $\quad\square$

Theorem 3.14. *Let A be finite dimensional. With the above notation the minimal projective resolution of the simple A_{loc}-module S_{loc} is of the form $P_i = P_{loc}^{x_i}$ with*

$$x_i = \sum_{x_p \in S_2(i)} o(k_p^2 + 2, v_{p-1}) + \sum_{x_p \in S_3(i)} o(2, v_p)$$

for $i \geq 2$, $x_1 = m_1$ and $x_0 = 1$. Moreover

$$\Omega^i(S_{loc}) = \bigoplus_{l \in Q_0} \bigoplus_{j=1}^{i} \Omega^j(S(l))_{loc}^{s_l^j}$$

for all $i \geq 1$ with

$$s_l^1 = \sum_{x_p \in S_3(i+1)} o_l(v_p) + \delta_{i1}$$

$$s_l^2 = \sum_{x_p \in S_3(i)} o_l(v_p) + \delta_{i2}$$

and for all $k \geq 3$ and $i \geq k - 1$

$$s_l^k = \sum_{x_p \in S_3(i-k+2)} o_l(v_p).$$

The symbol δ_{ij} is the Kronecker delta.

Proof. The exponents of the projective modules in the resolutions follow directly from Lemma 3.13. Let $x_p \in S_2(i)$ and consider the associated summand of $\Omega^i(S_{loc})$ by Lemma 3.13

$$O(k_p^2 + 2, v_{p-1}) = \bigoplus_{k_{t+1} \in Q_0} N(k_p^2 + 2, k_{t+1})^{o_{k_{t+1}}(v_{p-1})}$$

$$= \bigoplus_{k_{t+1} \in Q_0} \left[\Omega^{k_p^2+2}(S(k_{t+1}))_{loc}^{o_{k_{t+1}}(v_{p-1})} \oplus \right.$$

$$\left. \bigoplus_{j \in Q_0} \Omega^1(S(j))_{loc}^{(m_{k_p^2+2}^{k_{t+1}} - k_{t+1} m_{k_p^2+2}^j)} \right]^{o_{k_{t+1}}(v_{p-1})}$$

Hence we get a first summand of s_j^1 for all $j \in Q_0$ from the last big sum. The other sum yields a summand of $s_{k_{t+1}}^{k_p^2+2}$ for all $k_{t+1} \in Q_0$. We'll come back to that later. Now let $x_p \in S_3(i)$ and consider the associated summand of $\Omega^i(S_{loc})$

$$O(2, v_p)$$

$$= \bigoplus_{k_{t+1} \in Q_0} N(2, k_{t+1})^{o_{k_{t+1}}(v_p)}$$

$$= \bigoplus_{k_{t+1} \in Q_0} \left[\Omega^2(S(k_{t+1}))_{loc}^{o_{k_{t+1}}(v_p)} \oplus \bigoplus_{j \in Q_0} \Omega^1(S(j))_{loc}^{(m_1^{k_{t+1}} - k_{t+1} m_1^j)} \right]^{o_{k_{t+1}}(v_p)}$$

Again from the last big sum we get a second summand of s_j^1 for all $j \in Q_0$. The first big sum gives us the first summand of $s_{k_{t+1}}^2$ for all $k_{t+1} \in Q_0$.

For $i = 1$ we already know $\Omega^1(S_{loc}) = \bigoplus_{j \in Q_0} \Omega^1(S(j))_{loc}$. This yields the δ_{i1} in s_l^1 and the rest is zero by $S_2(1) = S_3(1) = \emptyset$. We get collectively for s_l^1

$$s_l^1 = \sum_{x_p \in S_2(i)} \sum_{k_{t+1} \in Q_0} o_{k_{t+1}}(v_{p-1})(m_{k_p^2+1}^{k_{t+1}} - k_{t+1} m_{k_p^2+1}^l)$$

$$+ \sum_{x_p \in S_3(i)} \sum_{k_{t+1} \in Q_0} o_{k_{t+1}}(v_p)(m_1^{k_{t+1}} - k_{t+1} m_1^l)$$

$$+ \delta_{i1}$$

$$= \sum_{x_p \in S_2(i)} o_l(v_{p-1}, k_p^2 + 1) + \sum_{x_p \in S_3(i)} o_l(v_p, 1) + \delta_{i1}$$

$$= \sum_{x_p \in S_3(i+1)} o_l(v_p) + \delta_{i1}.$$

Let's consider $s_{k_{t+1}}^{k_p^2+2}$ as above. The single sequence $x_p \in S_2(i)$ with $k_p^2 = 0$ is $x_0 = (1,1) \in S_2(2)$. In this case we get $o_{k_{t+1}}() = 1$ and therefore the summand δ_{i2} of $s_{k_{t+1}}^2$ for all $k_{t+1} \in Q_0$. For $i = 2$ the rest of s_l^2 is zero by $S_3(2) = \emptyset$.

44

At last we have to arrange the summands of $s_{k_t+1}^{k_p^2+2}$ from above by collecting all sequences $x_p \in S_2(i)$ that have the same k_p^2. We can write $x_p = (x_{p-1}, 2^{k_p^2})$ with $x_{p-1} \in S_3(i - k_p^2)$ belonging to $O(2, v_{p-1})$. This means we have for every $l \in Q_0$: $s_l^k = \sum_{x_{p-1} \in S_3(i-k+2)} o_l(v_{p-1})$.

We have considered all summands of $\Omega^i(S_{loc})$ for all $i \geq 1$ and collected the exponents in the s_l^j's. The assertion follows. $\square$

Corollary 3.15. *Let (P_i, d_i) be the minimal projective resolution of the simple A_{loc}-module S_{loc} of Theorem 3.14 with $P_i = P_{loc}^{x_i}$. Then the following is a minimal projective resolution of A_{loc} over A_{loc}^e:*

$$\ldots (P_{loc}^e)^{x_i} \xrightarrow{d_i \otimes 1} (P_{loc}^e)^{x_{i-1}} \to \ldots \to (P_{loc}^e)^{x_1} \xrightarrow{d_1 \otimes 1} P_{loc}^e \xrightarrow{d_0 \otimes 1} A_{loc} \to 0$$

Proof. Apply the exact functor $- \otimes_K A_{loc}^{op}$ to the resolution (P_i, d_i) of S_{loc}. We get $S_{loc} \otimes_K A_{loc}^{op} \simeq K \otimes_K A_{loc}^{op} \simeq A_{loc}$ and $P_{loc} \otimes_K A_{loc}^{op} \simeq A_{loc}^e \simeq P_{loc}^e$. $\square$

Remark 3.16. For local algebras there is an explicit interpretation of the complex $\text{Hom}_{A_{loc}^e}(P(A_{loc}), A_{loc})$ and of the Hochschild cohomology, where $P(A_{loc})$ is the projective resolution of A_{loc} from above.

Applying the functor $\text{Hom}_{A_{loc}^e}(-, A_{loc})$ to the resolution of A_{loc} we get

$$0 \to \text{End}_{A_{loc}^e}(A_{loc}) \to A_{loc} \xrightarrow{d_1^*} A_{loc}^{x_1} \xrightarrow{d_2^*} \ldots \to A_{loc}^{x_i} \xrightarrow{d_i^*} A_{loc}^{x_{i+1}} \to \ldots$$

where d_i^* is as follows: The maps $d_i : P_{loc}^{x_i} \to P_{loc}^{x_{i-1}}$ in the resolution of S_{loc} are A_{loc}-linear maps $d_i : A_{loc}^{x_i} \to A_{loc}^{x_{i-1}}$ and can be written as $(x_i \times x_{i-1})$-matrices with entries $d_i(j, k) \in A_{loc}$, $1 \leq j \leq x_{i-1}$ and $1 \leq k \leq x_i$,

$$d_i(a_1, \ldots, a_{x_i}) = d_i \cdot \begin{pmatrix} a_1 \\ \vdots \\ a_{x_i} \end{pmatrix}.$$

The algebra A_{loc}^e is local as well so we have again $P_{loc}^e = A_{loc}^e$. Thus the maps $d_i \otimes 1$ are the $(x_i \times x_{i-1})$-matrices with the entries $d_i(j, k) \otimes 1' \in A_{loc}^e$.

The isomorphism $\text{Hom}_{A_{loc}^e}((A_{loc}^e)^{x_i}, A_{loc}) \simeq A_{loc}^{x_i}$ is quiet easy. Usually we have $\text{Hom}_B(B, M) \to M$ with $f \mapsto f(1_B)$. In our case we can simply set $f \mapsto f \in A$, where the original map f was just left-multiplication with $f \in A$, the same with $f = (f_1, \ldots, f_{x_i}) \in \text{Hom}_{A_{loc}^e}((A_{loc}^e)^{x_i}, A) \simeq A^{x_i}$.

Let $f = (f_1, \ldots, f_{x_{i-1}}) \in A^{x_{i-1}}$ then

$$\operatorname{Hom}_{A^e_{loc}}(d_i \otimes 1, A_{loc})(f)$$
$$= f \circ (d_i \otimes 1)$$
$$= (f_1, \ldots, f_{x_{i-1}}) \cdot (d_i \otimes 1)$$
$$= (\sum_{j=1}^{x_{i-1}} f_j(d_i(j,1) \otimes 1), \ldots, \sum_{j=1}^{x_{i-1}} f_j(d_i(j,x_i) \otimes 1))$$
$$= (\sum_{j=1}^{x_{i-1}} f_j d_i(j,1), \ldots, \sum_{j=1}^{x_{i-1}} f_j d_i(j,x_i))$$
$$= f \cdot d_i.$$

Thus d_i^* is defined as

$$d_i^*(f) = f \cdot d_i.$$

It looks like $d_i^* = d_i^T$ but this is only true for commutative algebras! Finally we get $H^i(A_{loc}) = \ker d_{i+1}^* / \operatorname{Im} d_i^*$.

3.3 Splittable relations for monomial algebras

In this section we consider multiplication-relations and formulate conditions to detect them. Because these relations can be split to get a locally equivalent algebra in whose quiver these relations not exist, we will call them splittable relations. We have to distinguish between the monomial and the non-monomial case (section 4.1). In the first case we get a classification of the algebras into the classes locally finite/periodic/infinite (Theorem 3.23). In the latter case this is not so easy but we state the conditions for splittability (Lemma 4.1) because we can use it to construct the locally equivalent algebras.

Here are some direct observations how multiplication-relations behave. For instance, every loop α with $\alpha^2 \neq 0$ in the local algebra exists as a loop with nonzero square in any locally equivalent algebra.

Lemma 3.17. *Let $A_{loc} = KQ/I$ be a finite dimensional monomial local algebra, $Q_0 = \{1\}$, $Q_1 = \{\alpha_1, \ldots, \alpha_n\}$, let I be generated by the set of paths $S = \{x_1, \ldots, x_m\}$. Let $A = KQ'/I'$ be a finite dimensional monomial algebra that is locally equivalent to A_{loc}.*

(1) If there exists an arrow $\alpha_i \in Q_1$ with $\alpha_i^2 \notin I$, then the resolution of A_{loc} cannot be locally finite. If $\alpha_i^2 \notin I$ for all $i = 1, \ldots, n$, then the resolution of A_{loc} is locally infinite.

(2) In case $Q_1 = \{\alpha_1\}$, the resolution of A_{loc} is locally periodic if $\alpha_1^2 \notin I$ holds. Otherwise the resolution of A_{loc} is locally finite.

(3) Let A be connected and let l be the number of arrows $\alpha_i \in Q_1'$ with $\alpha_i^2 \in I$, then Q' has at most $l+1$ vertices.

(4) Let $n = 2$. If there exists an algebra A as above with $A \not\simeq A_{loc}$, then the resolution of A_{loc} must be locally finite or locally periodic.

(5) If there exists an infinite associated sequence of paths in A_{loc} which only consists of relations of the length > 2, then the global dimension of A is infinite.

Proof. (1) If the path $\alpha_i^2 \neq 0$ in A_{loc} does not vanish, the same holds for A, so the arrow must be a loop in A, too. Because I is monomial, there exists a relation $x_j = \alpha^l$ for some $l > 2$ and the same in I'. Thus there is a cycle in A that is covered by the generating relations. Because of Lemma 2.2 the resolution of A_{loc} cannot be locally finite. If this holds for all arrows, they must be loops in A as they are in A_{loc} and $A \simeq A_{loc}$.

(2) Let $\alpha_i^2 \in I$. There are no further relations that generate I because $I = KQ^{+2}$. This means for A that there is just one arrow, whose square is zero. So one possibility for A is the quiver with two vertices and one arrow between them together with $I' = \langle 0 \rangle$. This leads to gldim$A < \infty$ and A_{loc} is locally finite. If $\alpha_1^l = x_1$ is the single generating relation, then $\alpha_1^{l-1} \neq 0$ in A_{loc} like in A. So α_1 is a loop in A as well and $A_{loc} \simeq A$.

(3) There are at most l arrows in Q' that are no loops and they can at most connect $l+1$ vertices by a directed path.

(4) Let $A = KQ'/I'$ and $Q_1' = \{\alpha_1, \alpha_2\}$. According to Lemma 2.6, A has an infinite, non-periodic resolution if and only if there exist two cycles that are covered and connected by S with $I' = \langle S \rangle$. In case of two arrows this can only happen if both arrows are loops, which means A is local and $A \simeq A_{loc}$.

(5) By the construction of a local algebra the generating relations of I and I' differ only in relations of length 2. So every relation in the sequence is a generating relation for I' as well and the sequence is infinite in A. Hence gldim$A = \infty$ for all algebras A that are locally equivalent to A_{loc}. $\qquad\square$

Lemma 3.18. *Let $A = KQ/I$ be a finite dimensional monomial and connected algebra and let I be generated by the set of relations S. Let $1 \in Q_0$ and let $\alpha_1, \delta_1, \ldots, \delta_t$ be all arrows that start in 1 and let $\alpha_2, \gamma_1, \ldots, \gamma_n$ be all arrows that end in 1. Let the path $x = \alpha_1 \alpha_2$ be a relation, i.e., $x \in S$.*

Then there exists an algebra $B = KQ'/I'$ that is locally equivalent to A and in whose quiver Q' the path x does not exist if and only if there exists a basis (choice of arrows) of $\mathrm{rad}A/\mathrm{rad}^2 A$ such that there is no chain of paths between α_1 and α_2. This means that at least one path in every possible series of paths of the form

$$\alpha_1 \gamma_{j_1}, \ \delta_{i_1} \gamma_{j_1}, \ \delta_{i_1} \gamma_{j_2}, \ \ldots, \ \delta_{i_k} \gamma_{j_l}, \ \delta_{i_k} \alpha_2$$

is a relation in S.

A relation x that satisfies the condition of the lemma is called *a splittable relation* because in the quiver of the new algebra B the path x is split. This means x is a multiplication-relation in A. The algebra A is called *locally minimal* if there are no splittable relations in A.

Proof. Basic idea: The path x is zero in A. It is a multiplication-relation if in a new quiver Q' there is no path $\alpha_1\alpha_2$ because of $e(\alpha_2) \neq s(\alpha_1)$. So choose for Q' the starting point of α_1 as $s(\alpha_1) = 1$ (stays the same) and set $e(\alpha_2) = 2$ for a new vertex $2 \in Q'_0$. What happens to the other arrows δ_i and γ_j? It depends on which relations or paths in B additionally exist. Assume to a pair (δ_i, γ_j) the paths $\delta_i\gamma_j$, $\delta_i\alpha_2$, and $\alpha_1\gamma_j$ exist in B. This is the shortest possible chain of paths. Then they exist in every locally equivalent algebra. The quiver is not splittable as above because $2 = s(\delta_j) = e(\gamma_i) = 1$. This happens also for longer chains of paths because all arrows that occur in the chain belong to the same starting or ending point.

Let's do this explicitly. Assume there is a condition that gives us $s(\alpha_1) = e(\alpha_2)$ in Q. This happens if either there is a path $w \in A$ that has $x = \alpha_1\alpha_2$ as a subpath. But then $w = 0$ because x is a relation. Or if there is a chain of paths between α_1 and α_2, i.e., there is a series of paths of the form

$$\alpha_1\gamma_{j_1}, \delta_{i_1}\gamma_{j_1}, \delta_{i_1}\gamma_{j_2}, \ldots, \delta_{i_l}\gamma_{j_l}, \delta_{i_l}\alpha_2$$

and every path in that series is not zero in A. We get the condition

$$s(\alpha_1) = e(\gamma_{j_1}) = s(\delta_{i_1}) = e(\gamma_{j_2}) = \ldots = e(\gamma_{j_l}) = s(\delta_{i_l}) = e(\alpha_2).$$

We get: If there is no chain of paths between α_1 and α_2, we can build a new quiver Q' with $s(\alpha_1) = 1$ and $e(\alpha_2) = 2$. But what about the other arrows δ_i and γ_j that should be arranged on the vertices 1 and 2 in the new quiver as well? The same problem can happen here if there is an arrow δ_i (or γ_j) that is in relation to both α_1 and α_2 such that $s(\delta_i) = s(\alpha_1) = e(\alpha_2)$ (or $e(\gamma_j) = s(\alpha_1) = e(\alpha_2)$). This means that we have two chains of paths. The first between α_1 and δ_i (or γ_j) and the second between α_2 and δ_i (or γ_j). We can put these two chains together at the ends with δ_i or γ_j, respectively, and get a new chain of paths between α_1 and α_2.

We get: If there is no chain of paths between α_1 and α_2, we can build a new quiver Q' with $s(\alpha_1) = 1$ and $e(\alpha_2) = 2$ and every arrow δ_i and γ_j can be arranged consistently in the new quiver.

If on the other hand we have an algebra B in whose quiver Q' the path x does not exist and that is locally equivalent to A, there cannot be a chain of paths between α_1 and α_2. Because every path that exists in an algebra A exists in every locally equivalent algebra B as well. Thus the chain of paths would exist in B and would yield $s(\alpha_1) = e(\alpha_2)$, which is a contradiction. $\square$

Illustration. For a better understanding of splittable relations we consider some simple cases:

- Case $n = 1$, $t = 0$: The interesting part of Q is

$$s(\gamma_1) \xrightarrow{\gamma_1} 1 \xrightarrow{\alpha_1} e(\alpha_1) \quad s(\alpha_2) \xrightarrow{\alpha_2}$$

There is no path that connects γ_1 and α_2 (as for $t > 0$ the path $\delta_j\gamma_1$ if the path $\delta_j\alpha_2$ exists). So choose for Q'

$$s(\gamma_1) \xrightarrow{\gamma_1} 1 \xrightarrow{\alpha_1} e(\alpha_1) \qquad s(\alpha_2) \xrightarrow{\alpha_2} 2$$

Now x is a multiplication-relation, $I' = \langle S \setminus \{x\}\rangle$, $B' = KQ'/I'$ is locally equivalent to B. The case $k = 0$, $t = 1$ can be handled in a dual way, set $s(\delta_1) = 2$.

- Case $n = 1$, $t = 1$: Here the interesting part of Q looks like

$$s(\gamma_1) \xrightarrow{\gamma_1} 1 \xrightarrow{\delta_1} e(\delta_1) \quad s(\alpha_2) \xrightarrow{\alpha_2} 1 \xrightarrow{\alpha_1} e(\alpha_1)$$

In this case there is a single possible chain of paths between α_1 and α_2. The following table shows the possibilities to arrange γ_1 and δ_1 dependent on the existing paths. In doing so it is irrelevant if there exist relations of length > 2 that correspond to these paths because subpaths of generating relations are always nonzero. So the only important thing is which of the three paths occur as relations of length 2. We consider every case in the following table. In the first three columns of the table "1" means that the path exists, "0" that it does not exist which means it is a relation of length 2. In the last two columns "1" means that the particular arrow is placed at the vertex 1 in Q', "2" stays for the vertex 2 and "a" and "b" encode an arbitrary choice. If both columns contain "a", both arrows should be placed at the same vertex, which can be arbitrarily chosen.

$\delta_1\gamma_1$	$\delta_1\alpha_2$	$\alpha_1\gamma_1$	γ_1	δ_1
0	0	0	a	b
0	0	1	1	a
0	1	0	a	2
0	1	1	1	2
1	0	0	a	a
1	0	1	1	1
1	1	0	2	2

49

For example in the case $\delta_1\gamma_1 = 0$, $\delta_1\alpha_2 \neq 0$ and $\alpha_1\gamma_1 \neq 0$ we get the new quiver Q' with

$$s(\alpha_2) \xrightarrow{\alpha_2} 2 \xrightarrow{\delta_1} e(\delta_1)$$

$$s(\gamma_1) \xrightarrow{\gamma_1} 1 \xrightarrow{\alpha_1} e(\alpha_1)$$

So in every case the quiver is splittable and one gets an algebra that is locally equivalent to B and in that x is a multiplication-relation.

Remark 3.19. Let $A = KQ/I$ be a finite dimensional monomial algebra. Then there exists a locally minimal algebra which is locally equivalent to A. As we have seen in the above illustration, the splitting of the relations is not unique. So the locally minimal algebra is not unique in general. Compare the following example.

Example 3.20. Let Q be the following quiver

$$4 \xrightarrow{\alpha_2} 1 \underset{\alpha_1}{\overset{\delta}{<}} \begin{matrix} 3 \\ 2 \end{matrix}$$

and let I be the ideal of KQ generated by the paths $\delta\alpha_2$ and $\alpha_1\alpha_2$. Now there are two ways to split the relation $\alpha_1\alpha_2$ because the choice of $s(\delta)$ is arbitrary:

$$Q_1 \;=\; 4 \xrightarrow{\alpha_2} 5 \qquad 1 \underset{\alpha_1}{\overset{\delta}{<}} \begin{matrix} 3 \\ 2 \end{matrix}$$

$$Q_2 \;=\; 4 \xrightarrow{\alpha_2} 5 \xrightarrow{\delta} 3 \qquad 1 \xrightarrow{\alpha_1} 2$$

In case of Q_1 there is no relation left and $A_1 = KQ_1$ is locally minimal. In case of Q_2 there is still the relation $\delta\alpha_2$, which can be split as follows:

$$Q_3 = 4 \xrightarrow{\alpha_2} 5 \qquad 6 \xrightarrow{\delta} 3 \qquad 1 \xrightarrow{\alpha_1} 2$$

Now $A_3 = KQ_3$ is locally minimal but $A_3 \not\cong A_1$.

Corollary 3.21. *Let $A = KQ/I$ be a finite dimensional connected and monomial algebra. Let $(x_1, \ldots, x_n, \ldots)$ be an infinite associated sequence of paths to $x_1 \in S$ with $I = \langle S \rangle$ in A. There exists an algebra B that is locally equivalent to A and for which the sequence stops after a finite number of steps if and only if there is a splittable x_i in this sequence.*

Proof. Let $x_i \in S$ be a splittable relation that appears in the sequence the first time at position i. Then there exists an algebra B by Lemma 3.18 that is locally equivalent to A and in whose quiver the path x_i does not exist. An associated sequence of paths can only be built along a directed path in the quiver and with

50

the generating relations. There is no generator x_i of the ideal I' in $B = KQ'/I'$ anymore. The path along which the sequence in A was built stops in the middle of x_i in B so the sequence stops in B at the latest at position $i - 1$.

Now let B be locally equivalent to A and let the sequence in B stop at position $i - 1$. If the relation x_i would still exist in B, the sequence could be built at least one step further. So x_i cannot be a generating relation in B. Because A and B are locally equivalent, the path x_i must be zero in B as in A. By Lemma 3.18 x_i is splittable in A. □

Remark 3.22. Depending on the relations it happens often that a locally minimal algebra is disconnected. The extreme case is a quiver in which no arrow can be connected with another one. This happens if and only if the ideal I of the original algebra is generated by all paths of length 2. So for every finite quiver Q the resolution of the algebra KQ/KQ^{+2} is locally finite. But one can also find connected algebras whose resolutions have globally the properties that the resolution of an algebra A has locally. This can be seen in the proof of the following theorem.

Theorem 3.23. *Let $A = KQ/I$ be a finite dimensional monomial algebra.*

(1) *The resolution of A is locally finite if and only if each infinite associated sequence of paths contains at least one splittable relation.*

(2) *The resolution of A is locally periodic if and only if the number of infinite associated sequences of paths which contain no splittable relations is finite.*

(3) *The resolution of A is locally infinite if and only if the number of infinite associated sequences of paths which contain no splittable relations is infinite.*

Proof. Every infinite associated sequence of paths lives along at least one cycle of the quiver Q which is covered by S with $I = \langle S \rangle$. If there is a splittable relation at one vertex of the cycle, then the cycle is disconnected in a new algebra which was built like in the proof of Lemma 3.18. The sequence must stop like every sequence which lived along this cycle. In particular the new algebra is still connected because only a cycle was split at one vertex.

(1) Let the resolution of A be locally finite then there exists an algebra B which is locally equivalent to A and has a finite resolution. So there are no infinite sequences of paths in B. Hence by Corollary 3.21 all infinite sequences in A must contain a splittable relation. If on the other hand all sequences contain a splittable relation, there exists an algebra which is locally equivalent to A and in which all sequences are finite by the Corollary. So this algebra has finite global dimension.

(2) The resolution of A is locally periodic if and only if the resolution of A_{loc} is locally periodic. In A_{loc} there are infinitely many infinite associated sequences of paths. But an algebra with periodic resolution may just have finitely many infinite sequences. If B is locally equivalent to A and every infinite sequence of

paths in B has no splittable relations (exists by Lemma 3.18), then B has still finitely many infinite sequences and hence a periodic resolution if and only if A has finitely many infinite sequences without splittable relations.

(3) If A has infinitely many infinite sequences without splittable relations, then these sequences will be infinite in every locally equivalent algebra. So every to A locally equivalent algebra has an infinite, non-periodic resolution. $\qquad\square$

Example 3.24. All of the following algebras (compare Example 3.4) are in the same local equivalence class. All of their resolutions are locally finite:

$$A_1 \quad = \quad K\ 1 \xrightarrow{\alpha_1} 2 \xrightarrow{\alpha_2} 3 \xrightarrow{\alpha_3} 4 \ /\langle\alpha_3\alpha_2\alpha_1\rangle$$

$$A_2 \quad = \quad K\ 1 \xrightarrow{\alpha_1} 2 \underset{\alpha_3}{\overset{\alpha_2}{\rightleftarrows}} 3 \ /\langle\alpha_3\alpha_2\alpha_1, \alpha_2\alpha_3\rangle$$

$$A_3 \quad = \quad K\ 1 \xrightarrow{\alpha_1} 2 \overset{\alpha_3}{\underset{\alpha_2}{\circlearrowright}} \ /\langle\alpha_3\alpha_2\alpha_1, \alpha_2^2, \alpha_3^2, \alpha_2\alpha_3, \alpha_3\alpha_1\rangle$$

$$A_4 \quad = \quad K\ 1 \overset{\alpha_1}{\underset{\alpha_3}{\circlearrowright}} \alpha_2 \ /\langle\alpha_3\alpha_2\alpha_1, \alpha_1^2, \alpha_2^2, \alpha_3^2, \alpha_1\alpha_2, \alpha_1\alpha_3, \alpha_2\alpha_3, \alpha_3\alpha_1\rangle$$

One can read the algebras from bottom to top as the results of splitting relations. Splitting the relation α_1^2 in A_4 yields A_3, splitting $\alpha_3\alpha_1$ yields A_2. Finally splitting $\alpha_2\alpha_3$ leads to A_1 which is locally minimal and has finite global dimension. In this equivalence class there are some more algebras. Every quiver of them is of one of the following forms with the corresponding relations:

$$1 \xrightarrow{\alpha_1} 2 \xrightarrow{\alpha_2} 3 \overset{\alpha_3}{\circlearrowright} \ , \qquad 1 \xrightarrow{\alpha_1} 2 \overset{\alpha_2}{\circlearrowright} \xrightarrow{\alpha_3} 3 \ ,$$

$$\alpha_1 \circlearrowleft 1 \xrightarrow{\alpha_2} 2 \xrightarrow{\alpha_3} 3 \ , \qquad \alpha_1 \circlearrowleft 1 \overset{\alpha_2}{\circlearrowright} \xrightarrow{\alpha_3} 2 \ ,$$

$$\alpha_1 \circlearrowleft 1 \xrightarrow{\alpha_2} 2 \overset{\alpha_3}{\circlearrowright} \quad \text{and} \quad 1 \xrightarrow{\alpha_1} 2 \overset{\alpha_3}{\underset{\alpha_2}{\circlearrowright}} \ .$$

Example 3.25. The following algebra is locally minimal because there are no relations of length 2.

$$A = K\ 1 \xrightarrow{\alpha_1} 2 \ /\langle\alpha_3\alpha_2\alpha_1\alpha_3\rangle$$
$$\overset{\alpha_3}{\nwarrow} \underset{3}{} \overset{\alpha_2}{\nearrow}$$

It has a periodic resolution as one can see from the single relation $x = \alpha_3\alpha_2\alpha_1\alpha_3$, which can be overlapped infinitely many times with itself. Every subpath of x is nonzero in A. So in every algebra which is locally equivalent to A the following equations hold

$$\begin{aligned}
e(\alpha_3) &= s(\alpha_1) \\
e(\alpha_1) &= s(\alpha_2) \\
e(\alpha_2) &= s(\alpha_3).
\end{aligned}$$

We see that there are at most 3 vertices in a quiver of a locally equivalent algebra. If the quiver has exact 3 vertices, it must be the quiver of A and the algebra is A. If the quiver has only one vertex, it is the quiver of the local algebra

$$A_{loc} = K \; \alpha_1 \underset{\alpha_3}{\overset{\alpha_2}{\circlearrowright}} 1 \quad /\langle x, \alpha_1^2, \alpha_2^2, \alpha_3^2, \alpha_1\alpha_2, \alpha_2\alpha_3, \alpha_3\alpha_1\rangle.$$

Splitting any of the multiplication-relations of A_{loc}, we get the algebra $A_2 = KQ_2/I_2$ with

$$Q_2 = \alpha \circlearrowright 1 \underset{\gamma}{\overset{\beta}{\rightleftarrows}} 2$$

and $I_2 = \langle \alpha^2, \beta\gamma, \alpha\gamma\beta\alpha \rangle$. Splitting any of the remaining splittable relations, we get A again. So these three algebras build the whole local equivalence class of A.

Example 3.26. Let $A = KQ_2/I$ be the algebra with the quiver Q_2 from the above example and the ideal $I = \langle \alpha^3, \gamma\beta\gamma, \alpha\gamma \rangle$. The relations lead to the equations $s(\alpha) = e(\alpha) = e(\gamma) = s(\beta)$ and $s(\gamma) = e(\beta)$, which hold for every algebra B that is locally equivalent to A. Thus the quiver of B has at most two vertices; either both equations equal the same vertex, which yields the local algebra, or they equal different vertices, which yields A. So the local equivalence class consists only of A and

$$A_{loc} = K \; \gamma \underset{\alpha}{\overset{\beta}{\circlearrowright}} 1 \quad /\langle \alpha^3, \gamma\beta\gamma, \alpha\gamma, \gamma^2, \beta^2, \gamma\alpha, \alpha\beta\rangle.$$

Chapter 4

The non-monomial case

In the last two chapters, we saw many properties of the resolution of a monomial algebra. In the next step (section 4.2) we introduce the concept of an associated monomial algebra, which exists for every non-monomial algebra. We connect it to the idea of local equivalence. These two new algebras will help us to obtain properties of the resolution of the non-monomial algebra. For instance, it will turn out that the Anick/Green-resolution is minimal for A if and only if it is so for A_{loc} (Theorem 4.11). We get inequalities of the form: If the resolution of A_{mon} is locally periodic, then the resolution of A is either locally infinite or locally periodic (Theorem 4.15).

Throughout this chapter let $A = KQ/I$ be a finite dimensional K-algebra with a suitable ordering $<$ on the set B of paths of the quiver Q. Let the ideal I be generated by the set S which need not consist exclusively of paths; some generators may be linear combinations of paths. Let A_{loc} denote the local algebra which is locally equivalent to A.

4.1 Splittable relations for non-monomial algebras

There is an analogue of Lemma 3.18 in the non-monomial case. The difference lies only in those generating relations that are linear combinations of more than one path. This does not help us to understand the resolution because we cannot use the resolution of Bardzell. But again we can construct every locally equivalent algebra by splitting the splittable relations.

Notation. We extend the term of a chain of paths of Lemma 3.18 for non-monomial algebras. Let $A = KQ/I$ be a finite dimensional non-monomial and connected algebra and let I be generated by the set of relations S, that are all uniform. Let $x = \delta_1 \gamma_1$ be a monomial relation in S with inner vertex $1 \in Q_0$. Let $\delta_2, \ldots, \delta_t$ be all further arrows that start in 1 and let $\gamma_2, \ldots, \gamma_n$ be all further arrows that end in 1.

An *extended chain of paths* between two of the arrows on the vertex 1 is an iteration of the following where one of the two arrows is at the start and the other one is at the end of the iteration:

- The standard chain of paths between δ_i and δ_j, $i, j \in \{1, \ldots, t\}$,

$$\delta_i \gamma_{j_1}, \; \delta_{i_1} \gamma_{j_1}, \; \delta_{i_1} \gamma_{j_2}, \; \ldots, \; \delta_{i_m} \gamma_{j_m}, \; \delta_j \gamma_{j_m}.$$

  This means every path in the chain exists in A thus $s(\delta_i) = s(\delta_j)$ in every algebra in the local equivalence class of A.

- The standard chain of paths between δ_i and γ_j, $i \in \{1, \ldots, t\}$, $j \in \{1, \ldots, n\}$,

$$\delta_i \gamma_{j_1}, \; \delta_{i_1} \gamma_{j_1}, \; \delta_{i_1} \gamma_{j_2}, \; \ldots, \; \delta_{i_m} \gamma_{j_m}, \; \delta_{i_m} \gamma_j.$$

  This means every path in the chain exists in A thus $s(\delta_i) = e(\gamma_j)$ in every algebra in the local equivalence class of A.

- The standard chain of paths between γ_i and γ_j, $i, j \in \{1, \ldots, n\}$,

$$\delta_{i_1} \gamma_i, \; \delta_{i_1} \gamma_{j_1}, \; \delta_{i_2} \gamma_{j_1}, \; \ldots, \; \delta_{i_m} \gamma_{j_m}, \; \delta_{i_m} \gamma_j.$$

  This means every path in the chain exists in A thus $e(\gamma_i) = e(\gamma_j)$ in every algebra in the local equivalence class of A.

- The *commutativity relations* $y = \sum k_l w_l \in S$ between δ_i and δ_j, $i, j \in \{1, \ldots, t\}$, with $k_l \in K$ and $w_l \in B$ such that δ_i is the first arrow of some w_i and δ_j is the first arrow of some w_j. This implies again $s(\delta_i) = s(\delta_j)$.

- The *commutativity relations* $y = \sum k_l w_l \in S$ between γ_i and γ_j, $i, j \in \{1, \ldots, n\}$, with $k_l \in K$ and $w_l \in B$ such that γ_i is the last arrow of some w_i and γ_j is the last arrow of some w_j. This implies again $e(\gamma_i) = e(\gamma_j)$.

This means an extended chain of paths is an alternating sequence of standard chains of paths and commutativity relations between some arrows.

Lemma 4.1. *Let $A = KQ/I$ be a finite dimensional non-monomial and connected algebra. Let I be generated by the set of uniform relations S. Let $x = \delta_1 \gamma_1$ be a monomial relation in S.*

Then there exists an algebra $A' = KQ'/I'$ that is locally equivalent to A and in whose quiver Q' the path x does not exist if and only if there is a basis (choice of arrows) of $\mathrm{rad} A / \mathrm{rad}^2 A$ such that there is no extended chain of paths between δ_1 and γ_1.

If the condition of the lemma is satisfied for $x \in S$, then x is again called *splittable*.

Proof. The relation x is splittable if and only if we can build a quiver Q' in which the path $x = \delta_1 \gamma_1$ does not exist because $s(\delta_1) \neq e(\gamma_1)$ in Q'. This means we set

$$s(\gamma_1) \xrightarrow{\gamma_1} 2 \qquad 1 \xrightarrow{\delta_1} e(\delta_1)$$

in Q' with a new vertex 2. We arrange all the other arrows δ_i and γ_j on the vertices 1 and 2 such that every condition like $e(\gamma_j) = s(\delta_i)$ that concerns the former vertex 1 is still satisfied. Every other vertex in Q' and every other arrow is still as in Q. This means $A' = KQ'/I'$ is locally equivalent to A where I' is the ideal generated by all generators of I except the paths that were split.

This splitting is possible if and only if there is no condition that finally implies $s(\delta_1) = e(\gamma_1)$ even in Q'. This is equivalent to: There is no extended chain of paths between δ_1 and γ_1 (compare the proof of Lemma 3.18). In the non-monomial case here we have the commutativity relations as a speciality. We want them to be uniform (otherwise decompose them into uniform ones) thus every path in the relation has the same starting and ending point. If we don't have a standard chain of paths between δ_1 and γ_1, there can still be an extended chain of paths that contradicts our construction of Q'. By assumption there is no such chain and we are done. $\square$

Example 4.2. Consider the algebra A in Example 1.23. It is not locally minimal because the relation $\alpha_4\alpha_3$ is splittable to

$$
Q' = \quad
\begin{array}{c}
2 \\
{\scriptstyle\alpha_1}\nearrow \quad \searrow {\scriptstyle\alpha_2} \\
1 \xrightarrow{\ \beta\ } 3 \\
{\scriptstyle\alpha_4}\uparrow \quad \swarrow {\scriptstyle\alpha_3} \\
4 \qquad 5
\end{array}
$$

With $S' = \{\alpha_1\alpha_4, \, \alpha_3\beta - \alpha_3\alpha_2\alpha_1\}$ we get the locally minimal algebra $A' = KQ'/\langle S'\rangle$. The relation $\alpha_1\alpha_4$ is not splittable because we have a chain of paths between α_1 and α_4 by $y = \alpha_3\beta - \alpha_3\alpha_2\alpha_1$ and $\beta\alpha_4 \neq 0$. The resolutions of the simple A'-modules are finite. Thus the resolution of A' is finite and the resolution of A is locally finite.

4.2 The associated monomial algebra

To each non-monomial algebra there exists at least one associated monomial algebra. We will show that the Anick/Green-resolution of this monomial algebra and the Anick/Green-resolution of the non-monomial algebra are closely connected.

Definition 4.3. Let $A = KQ/I$ be a non-monomial finite dimensional K-algebra. Let $<$ be a suitable ordering on the set of paths B of Q and let Q_2 be the obstruction set (defined in 1.3) dependent on the surjection $f : KQ \to A$. Then $A_{mon} = KQ/\langle Q_2\rangle$ is called the *associated monomial algebra* of A.

Lemma 4.4. *The monomial ideal generated by Q_2 is equal to the monomial ideal generated by* $\mathrm{Tip}(I)$.

Proof. Recall

$$M = \{\beta \in B \mid f(\beta) \notin \mathrm{span}(f(\gamma) \mid \gamma < \beta)\},$$
$$Q_2 = \{\beta \in B \setminus M \mid \text{whenever } \beta = \beta_1\beta_2 \text{ with } \beta_1, \beta_2 \in B \setminus Q_0,$$
$$\text{then } \beta_1 \in M \text{ and } \beta_2 \in M\},$$
$$\mathrm{Tip}(I) = \{x \in B \mid x = \mathrm{tip}(y), \ y \in I\}.$$

First let $x \in Q_2$. Then $x \in B \setminus M$ and $f(x) \in \mathrm{span}(f(\gamma) \mid \gamma < x)$. So let $f(x)$ be equal to $\sum_i \alpha_i f(y_i)$ for $\alpha_i \in K$ and $y_i < x$ for all i. Then $x - \sum_i \alpha_i y_i \in \ker f = I$. Because $x > y_i$ for all i, $x = \mathrm{tip}(x - \sum_i \alpha_i y_i)$ thus x is in $\mathrm{Tip}(I)$.

Now let x be an element of $\mathrm{Tip}(I)$. Then there exists an element $y = \sum_{i=1}^k \alpha_i y_i \in I$ with $x = y_1 = \mathrm{tip}(y)$. This implies $y_i < x$ for all $i \geq 2$ and $f(x) = -\frac{1}{\alpha_1} f(y - \alpha_1 x) \in \mathrm{span}(f(y_i) \mid i \geq 2)$ so $x \notin M$. If x is in Q_2, we are done. Otherwise x can be written as $x = \beta_1\beta_2$ with $\beta_1, \beta_2 \in B \setminus Q_0$ and $\beta_1 \notin M$ or $\beta_2 \notin M$. If β_1 is not in M, then $f(\beta_1) = \sum_i \alpha_i f(\gamma_i)$ and like above $\beta_1 \in \mathrm{Tip}(I)$, the same with β_2. Now one can do this again with β_1 or β_2 as with x. Finally there is a subpath of x which lies in Q_2 because all arrows are in M. Hence at the latest a subpath of x of length two lies in Q_2. $\square$

Lemma 4.5. *Let $A = KQ/I$ be non-monomial and let A_{mon} be the associated monomial algebra (dependent on $<$ and f). Let (P^i_{mon}, d^i_{mon}) denote the Anick/Green-resolution of a simple A_{mon}-module S_{mon}, $P^i_{mon} = \bigoplus_{j \in Q_0} P^{n^i_j}_{mon}(j)$ for $n^i_j \geq 0$.*

Then the Anick/Green-resolution of the simple A-module S that belongs to the same vertex as S_{mon} consists of the projective modules $P^i = \bigoplus_{j \in Q_0} P^{n^i_j}(j)$ and of the associated morphisms d^i.

Proof. According to the construction the following sets are the same for A as for A_{mon}: B as a K-basis of KQ, Q_2 (see Corollary 1.22) and for this reason $M = \{\beta \in B \mid \text{no subpath of } \beta \text{ lies in } Q_2\}$. The construction of the sets Q_m does not depend on the ideal I nor on $f : KQ \rightarrow A$ or $f_{mon} : KQ \rightarrow A_{mon}$. So the sets Q_m are the same for A as for A_{mon}. Both resolutions consist (Theorem 1.21) of the modules $K_m = A \otimes_R KQ^i_m$ and $(K_{mon})_m = A_{mon} \otimes_{R_{mon}} KQ^i_m$, respectively. Let $Q^i_m = \{w_1, \ldots, w_l\}$ then $K_m = \bigoplus_{j=1}^l P(e(w_j))$ and $(K_{mon})_m = \bigoplus_{j=1}^l P_{mon}(e(w_j))$. $\square$

To compute the Anick/Green-resolution one can always use Bardzell's resolution in the version for simple modules. Either the algebra itself is monomial or the Anick/Green-resolution equals the minimal resolution of the associated monomial algebra, which is Bardzell's resolution.

The resolution of Anick/Green is often not minimal for non-monomial algebras, it depends strongly on the choice of the order.

If for an algebra A there exists a suitable order and an associated monomial algebra A_{mon} such that the Anick/Green-resolution dependent on the order is minimal for each simple A-module, we say that (in general) *the Anick/Green-resolution is minimal for A*. The resolution is not minimal for A if there is no suitable order such that Anick/Green is minimal for each simple A-module.

For example the Anick/Green-resolution is not minimal for the algebra A of Example 1.23 because the obstruction set is the same for every suitable order.

4.3 Connections to local algebras

Lemma 4.6. $(A_{loc})_{mon} = (A_{mon})_{loc}$

Notation. Let $A = KQ/I$ be as above and let $\alpha \in Q_1$ be an arrow in Q. Then α^* denotes the associated arrow in Q_{loc}. If $w = \alpha_1 \ldots \alpha_n$ is a path in Q, let $w^* = \alpha_1^* \ldots \alpha_n^*$ denote the associated path in Q_{loc}. Let $X = \{w_1, \ldots, w_m\}$ be a set of paths in Q, then $X^* = \{w_1^*, \ldots, w_m^*\}$ is the associated set of paths in Q_{loc}. Let $<$ denote the suitable order on the set of paths A of KQ then $<_{loc}$ is the suitable order on A_{loc} with $w_1^* <_{loc} w_2^*$ iff $w_1 < w_2$ for $w_1, w_2 \in KQ$ and a reasonable continuation on the multiplication-relations in KQ_{loc}.

Proof. By definition and Lemma 4.4 the associated monomial algebra to A is

$$A_{mon} = KQ/\langle \mathrm{Tip}(I) \rangle.$$

Let I be generated by a minimal set S of linear combinations of paths. Let $Y = \{\alpha_i \alpha_j \mid \alpha_i, \alpha_j \in Q_1, s(\alpha_i) \neq e(\alpha_j)\}$ be the set of zero multiplications in KQ. Then $I_{loc} = \langle S^*, Y^* \rangle$. And it is easy to see that $\mathrm{Tip}(I_{loc}) = (\mathrm{Tip}(I))^* \cup \langle Y^* \rangle$ because Y is a set of paths. Now we have

$$\begin{aligned}
(A_{loc})_{mon} &= KQ_{loc}/\langle \mathrm{Tip}(I_{loc}) \rangle \\
&= KQ_{loc}/\langle (\mathrm{Tip}(I))^*, Y^* \rangle.
\end{aligned}$$

On the other hand

$$\begin{aligned}
(A_{mon})_{loc} &= (KQ/\langle \mathrm{Tip}(I) \rangle)_{loc} \\
&= KQ_{loc}/\langle \mathrm{Tip}(I) \rangle_{loc}
\end{aligned}$$

where $\langle \mathrm{Tip}(I) \rangle_{loc}$ is generated like above (compare I_{loc}) by the associated set of the original generators $(\mathrm{Tip}(I))^*$ and the associated set of multiplication-relations Y^*. $\qquad \square$

Corollary 4.7. *If $[A]$ denotes the local equivalence class of an algebra A, then $B \in [A]$ implies $B_{mon} \in [A_{mon}]$.*

Example 4.8. Unfortunately the converse is not true. Take for example the algebra

$$A = K \; 1 \underset{v_1}{\overset{u_1}{\rightleftarrows}} 2 \underset{v_2}{\overset{u_2}{\rightleftarrows}} 3 \quad /\langle u_2 u_1, \; v_1 v_2, \; v_2 u_2 - u_1 v_1 \rangle.$$

This is a simple example for a Brauer tree algebra. It is biserial and has a periodic resolution. The relations $u_2 u_1$ and $v_1 v_2$ are both not splittable because of the commutativity relation. The associated monomial algebra has the same quiver but the ideal is generated by $u_2 u_1, \; v_1 v_2, \; v_2 u_2, \; u_1 v_1 u_1, \; v_1, u_1 v_1$. Thus we can split some relations and get finally the locally minimal algebra

$$A_{min} = K \; (\; 1 \underset{v_1}{\overset{u_1}{\rightleftarrows}} 2 \qquad 3 \xrightarrow{v_2} 4 \xrightarrow{u_2} 5 \;) \quad /\langle u_1 v_1 u_1, \; v_1 u_1 v_1 \rangle.$$

There is no $B \in [A]$ such that $B_{mon} \simeq A_{min}$.

This means we have to introduce a smaller version of the local equivalence class of A_{mon}:

Definition 4.9. Let $[A]_{mon}$ denote the set of all $B_{mon} \in [A_{mon}]$ with $B \in [A]$. We say that the resolution of A_{mon} or of its simple modules is *locally quasi finite/periodic/infinite* if there is a $B_{mon} \in [A]_{mon}$ which has a finite/periodic/infinite resolution and no algebra with a resolution of a "smaller" type.

As a direct consequence we get the following:

Lemma 4.10. *Let A_{mon} be the associated monomial algebra to A.*

(1) If the resolution of A_{mon} is locally quasi finite, then it is also locally finite.

(2) If the resolution of A_{mon} is locally quasi periodic, then it is either locally finite or locally periodic.

The same holds for the resolution of the simple A_{mon}-modules.

Proof. If the resolution of A_{mon} is locally quasi finite, then there is an algebra $B_{mon} \in [A]_{mon}$ with a finite resolution. Because $B_{mon} \in [A_{mon}]$, the resolution of A_{mon} is locally finite. If the resolution of A_{mon} is locally quasi periodic, then there is a locally equivalent algebra with a periodic resolution. In $[A]_{mon}$ there is no algebra with a finite resolution. But in $[A_{mon}]$ this is still possible. So the resolution of A_{mon} is either locally finite or locally periodic. $\square$

To analyse the question why the Anick/Green-resolution is sometimes minimal for non-monomial algebras and sometimes not, it is enough to consider only local algebras. The property that the resolution is minimal is invariant under local equivalence:

Theorem 4.11. *Let $A = KQ/I$ be a finite dimensional K-algebra and let $<$ be a suitable order on the K-basis of the path algebra KQ. Let A_{loc} be the local algebra that is locally equivalent to A. Then the Anick/Green-resolution for A associated to $<$ is minimal if and only if the Anick/Green-resolution for A_{loc} associated to $<_{loc}$ is minimal.*

Proof. By Theorem 3.14 we know that there are integers s_l^j such that

$$\Omega^i(S_{loc}) = \bigoplus_{l \in Q_0} \bigoplus_{j=1}^i \Omega^j(S(l))_{loc}^{s_l^j}$$

for all $i \geq 1$. In A-mod the following sequence, which is the projective cover of $\Omega^j(S(l))$ and its kernel, is exact:

$$0 \to \Omega^{j+1}(S(l)) \to \bigoplus_{k \in Q_0} P(k)^{{}^l m_j^k} \to \Omega^j(S(l)) \to 0 \qquad (4.1)$$

The projective cover of $\Omega^j(S(l))_{loc}$ in A_{loc}-mod is

$$0 \to \Omega^{j+1}(S(l))_{loc} \oplus \bigoplus_{k \in Q_0} \bigoplus_{m \neq k} \Omega^1(S(m))_{loc}^{{}^l m_j^k} \to P_{loc}^{m_j^l} \to \Omega^j(S(l))_{loc} \to 0 \qquad (4.2)$$

where $m_j^l = \sum_{k \in Q_0} {}^l m_j^k$ as in 3.2.3. We see that the minimal resolution of S_{loc} is closely connected to the minimal resolutions of all simple modules $S(j) \in A$-mod and their images under $-_{loc}$ in A_{loc}-mod.

Let the Anick/Green-resolution be minimal for A_{loc}. This means the structure of the minimal resolution of S_{loc} is the same as the structure of the minimal resolution of the simple module S_{loc}^{mon} of the associated monomial algebra A_{loc}^{mon} (Lemma 4.5). Thus

$$\Omega^i(S_{loc}) = \bigoplus_{l \in Q_0} \bigoplus_{j=1}^i \Omega^j(S(l))_{loc}^{s_l^j}$$

and

$$\Omega^i(S_{loc}^{mon}) = \bigoplus_{l \in Q_0} \bigoplus_{j=1}^i \Omega^j(S(l)^{mon})_{loc}^{s_l^j}$$

for the same integers s_l^j in both equalities. Because we have a similar version of (4.2) in A_{loc}^{mon}-mod where the modules are associated to the same vertices and have the same multiplicities, it follows that

$$0 \to \Omega^{i+1}(S(j)^{mon}) \to \bigoplus_{k \in Q_0} (P(k)^{mon})^{{}^l m_j^k} \to \Omega^i(S(j)^{mon}) \to 0$$

61

is exact in A_{mon}-mod. It has the same structure as (4.1) in A-mod for all i. Hence the Anick/Green-resolution is minimal for A.

Assuming the Anick/Green-resolution is minimal for A means that the resolutions of $S(j)$ in A-mod and $S(j)^{mon}$ in A_{mon}-mod have the same structure like above for all j. This yields the same structure for the local resolutions and hence the assertion. □

4.4   The resolution for non-monomial algebras

In this section we connect all that we know about the resolution of a non-monomial algebra. We compare the resolution of the algebra with the resolutions of the simple modules with respect to the local algebra. At last we compare the resolution of the algebra with the resolution of the associated monomial algebra. Hence in many cases it is enough to compute the minimal resolution of A_{mon} to get information about the resolution of A.

Lemma 4.12. *Let A be finite dimensional. Let the Anick/Green-resolution be minimal for A. If the resolution of A is almost periodic, then the resolutions of the simple A-modules are periodic.*

By Remark 2.4 this is equivalent to: Let the Anick/Green-resolution be minimal. The resolutions of the simple A-modules are almost periodic if and only if they are periodic.

Proof. Let the resolution of A be almost periodic. Then the resolutions of the simple modules are almost periodic by Lemma 1.3. The Anick/Green-resolution for A is minimal thus this resolution is also almost periodic. It has the same structure as the minimal resolutions of the simple A_{mon}-modules. So the resolutions of the simple A_{mon}-modules are almost periodic, too. By Lemma 2.7 they are periodic because A_{mon} is monomial. Hence the resolutions of the simple A-modules are periodic. □

Proposition 4.13. *Let A be finite dimensional.*

(1) The resolution of A is locally finite if and only if the resolutions of the simple A-modules are locally finite.

(2) The resolution of A is locally almost periodic if and only if the resolutions of the simple A-modules are locally almost periodic.

(3) The resolution of A is locally infinite if and only if the resolutions of the simple A-modules are locally infinite.

(4) Let the Anick/Green-resolution be minimal for A. If the resolution of A is locally periodic, then the resolutions of the simple A-modules are locally periodic.

The same hold for locally quasi finite/periodic/infinite.

Proof. (1) By definition.

(2) Let the resolution of A be locally almost periodic. Then by definition there is an algebra $B \in [A]$ which has an almost eventually periodic resolution. By Lemma 1.3 and (1.1) every simple B-module has an almost eventually periodic resolution. There is no algebra in the local equivalence class with finite global dimension. Thus the resolutions of the simple A-modules are locally almost periodic. The other direction uses the same arguments.

(3) Let the resolution of A be locally infinite. Then every algebra $B \in [A]$ has an infinite, non-periodic resolution. This means the number of projective modules in the resolution grows with the level of the resolution. Thus with Lemma 1.3 and (1.1) the resolution of the simple B-modules are infinite and not periodic. Then the resolutions of the simple A-modules are locally infinite. The other direction uses the same arguments.

(4) Assume now the Anick/Green-resolution is minimal for A and the resolution of A is locally periodic. Then there is an algebra $B \in [A]$ with a periodic resolution. By Lemma 1.3 together with (1.1) the resolutions of the simple B-modules are almost periodic. Thus they are periodic by Lemma 4.12 and the resolutions of the simple A-modules are locally periodic. $\square$

Proposition 4.14. *Let A be finite dimensional.*

(1) *If the resolutions of the simple A_{mon}-modules are locally quasi finite, then the resolutions of the simple A-modules are locally finite.*

(2) *If the resolutions of the simple A-modules are locally almost periodic, then the resolutions of the simple A_{mon}-modules are locally quasi periodic or locally quasi infinite.*

(3) *If the resolutions of the simple A_{mon}-modules are locally quasi periodic, then the resolutions of the simple A-modules are locally almost periodic or locally finite.*

(4) *If the resolutions of the simple A-modules are locally infinite, then the resolutions of the simple A_{mon}-modules are locally quasi infinite.*

Let the Anick/Green-resolution be minimal for A.

(5) *If the resolutions of the simple A-modules are locally finite, then the resolutions of the simple A_{mon}-modules are locally finite.*

(6) *The resolutions of the simple A-modules are locally almost periodic if and only if the resolutions of the simple A_{mon}-modules are locally quasi periodic.*

(7) *If the resolutions of the simple A_{mon}-modules are locally infinite, then the resolutions of the simple A-modules are locally infinite.*

Proof. (1) Let the resolutions of the simple A_{mon}-modules be locally quasi finite. Then there is an algebra $B_{mon} \in [A]_{mon}$ with finite global dimension. Because the Anick/Green-resolution of the simple B_{mon}-modules has the same structure as the Anick/Green-resolution of the simple B-modules and is finite, the minimal

63

resolutions of the simple B-modules must be finite, too. So the resolutions of the simple A-modules are locally finite.

(2) Let the resolutions of the simple A-modules be locally almost periodic. Then there is an algebra $B \in [A]$ whose simple modules have almost periodic resolutions. Either the Anick/Green-resolution is minimal for B, then the resolutions of the simple B_{mon}-modules are periodic. Or the resolution is not minimal and the resolutions of the simple B_{mon}-modules are infinite and not periodic. Thus the resolutions of the simple A_{mon}-modules are either locally quasi periodic or locally quasi infinite.

(3) Let the resolutions of the simple A_{mon}-modules be locally quasi periodic. Then there is an algebra $B_{mon} \in [A]_{mon}$ whose simple modules have periodic resolutions. The Anick/Green-resolutions of the simple B-modules have the same structure as the minimal resolutions of the simple B_{mon}-modules, which are periodic. They are upper bounds for the minimal resolutions of the simple B-modules. Thus the resolutions of the simple B-modules are almost periodic or finite. Hence the resolutions of the simple A-modules are locally almost periodic or locally finite.

(4) Let the resolutions of the simple A-modules be locally infinite. Then the resolutions of the simple modules of every algebra $B \in [A]$ are infinite and not periodic. The Anick/Green-resolution as an upper bound must be infinite and non-periodic, too. Thus the resolutions of the simple B_{mon}-modules are infinite and non-periodic and the resolutions of the simple A_{mon}-modules are locally quasi infinite.

Let the Anick/Green-resolution be minimal for every simple A-module.

(5) Let the resolution of the simple A-modules be locally finite. Then there is a $B \in [A]$ whose simple modules have all a finite resolution. Here the minimal resolution is the Anick/Green-resolution. It has the same structure as the resolution of the simple B_{mon}-modules. So the resolution of the simple A_{mon}-modules are locally finite.

(6) ($\Rightarrow$) Let the resolutions of the simple A-modules be locally almost periodic. Then there is a $B \in [A]$ whose simple modules have almost periodic resolutions. Because the Anick/Green-resolution is minimal, the resolutions of the simple B_{mon}-modules are periodic. Hence the resolutions of the simple A_{mon}-modules are locally periodic or locally finite. They are locally finite if there is a locally equivalent algebra with finite global dimension which has no correspondent in $[A]$. This means the resolutions are locally quasi periodic.

($\Leftarrow$) Let the resolutions of the simple A_{mon}-modules be locally quasi periodic. Then there is an algebra $B_{mon} \in [A]_{mon}$ with a periodic resolution but no algebra with a finite resolution. Because the Anick/Green-resolution is minimal, the resolution of the corresponding $B \in [A]$ is periodic. So the resolutions of the simple A-modules are locally periodic.

(7) Let the resolutions of the simple A_{mon}-modules be locally infinite. This means there is no locally equivalent algebra with a periodic or finite resolution;

64

neither in $[A]_{mon}$ nor in $[A_{mon}]$. Because the Anick/Green-resolution is minimal, every $B \in [A]$ has at least one simple module with an infinite, non-periodic resolution. Thus the resolutions of the simple A-modules are locally infinite. $\square$

Theorem 4.15. *Let A be finite dimensional. The following hold:*

(1) *If the resolution of A_{mon} is locally quasi finite, then the resolution of A is locally finite.*

(2) *If the resolution of A is locally almost periodic, then the resolution of A_{mon} is either locally quasi periodic or locally quasi infinite.*

(3) *If the resolution of A_{mon} is locally quasi periodic, then the resolution of A is either locally finite or locally almost periodic.*

(4) *If the resolution of A is locally infinite, then the resolution of A_{mon} is locally quasi infinite.*

If the resolution of Anick and Green is minimal for A, we get:

(5) *If the resolution of A is locally finite, then the resolution of A_{mon} is locally finite.*

(6) *The resolution of A is locally almost periodic if and only if the resolution of A_{mon} is locally quasi periodic.*

(7) *If the resolution of A_{mon} is locally infinite, then the resolution of A is locally infinite.*

Proof. Follows by a combination of Propositions 4.13 and 4.14. For instance consider (1). Let the resolution of A_{mon} be locally quasi finite. Then by Proposition 4.13 (1) the resolutions of the simple A_{mon}-modules are locally quasi finite. This implies (Proposition 4.14 (1)) that the resolutions of the simple A-modules are locally finite. Again with Proposition 4.13 (1) we get that the resolution of A is locally finite.

Alternatively we can prove the Theorem without the propositions as follows:

(1) Let the resolution of A_{mon} be locally quasi finite. Then there exists an algebra $B_{mon} \in [A]_{mon}$ which has a finite resolution. Thus by Happel's Lemma 1.3 or by Corollary 1.17 the resolutions of the simple B_{mon}-modules are finite. The Anick/Green-resolutions of the simple B-modules are also finite by Lemma 4.5. They are an upper bound for the minimal resolutions of the simple B-modules. Hence the resolutions of the simple B-modules are also finite. Again with Happel, the resolution of B is finite and thus the resolution of A is locally finite.

(2) Let the resolution of A be locally almost periodic. This means there is an algebra $B \in [A]$ with an almost periodic resolution but none with a finite resolution. With Happel, the resolutions of the simple B-modules are almost periodic. The Anick/Green-resolution as an upper bound of the minimal resolution of a simple B-module must be almost periodic or infinite and non-periodic. It has the same structure as the Anick/Green-resolution for the associated simple B_{mon}-module. Hence the resolution of B_{mon} is either almost periodic or infinite.

Because B_{mon} is monomial, the resolution is either periodic or infinite. By (1) there is no locally equivalent algebra with a finite resolution in $[A]_{mon}$. Hence the resolution of A_{mon} is locally quasi periodic or locally quasi infinite.

(3) Let A_{mon} be locally quasi periodic. Then there is an algebra $B_{mon} \in [A]_{mon}$ with a periodic resolution. The resolutions of the simple B_{mon}-modules are also periodic. With the same arguments as above, the resolutions of the simple B-modules are almost periodic or finite. By Happel, the resolution of B is almost periodic or finite. Hence the resolution of A is locally almost periodic or locally finite.

(4) Let the resolution of A be locally infinite. Then every locally equivalent algebra B has an infinite, non-periodic resolution. By Happel, its simple modules have infinite, non-periodic resolutions, too. Because the Anick/Green-resolution is an upper bound, it must be infinite and non-periodic, too. The same for the resolutions of the simple B_{mon}-modules. Hence the resolution of B_{mon} is infinite and non-periodic and the resolution of A_{mon} is locally quasi infinite.

From now on let the Anick/Green-resolution be minimal for A.

(5) Let the resolution of A be locally finite. Then there is a $B \in [A]$ with a finite resolution. By Happel, the simple B-modules have finite resolutions. In particular, the Anick/Green-resolutions are finite because they are minimal for B by Theorem 4.11. They have the same structure as the Anick/Green-resolutions of the simple B_{mon}-modules. Thus the resolution of B_{mon} is finite and the resolution of A_{mon} is locally finite.

(6) ($\Rightarrow$) Let the resolution of A be locally almost periodic. Then we have $B \in [A]$ with an almost periodic resolution. This means by Happel the resolutions of the simple B-modules are almost periodic. The Anick/Green-resolution is minimal for B, hence the resolutions of the simple B_{mon}-modules are periodic. The same holds for the resolutions of the simple B-modules. It follows that the resolution of B_{mon} is periodic and the resolution of A_{mon} is locally quasi periodic.

($\Leftarrow$) Uses the same arguments. Only mention that Happel reduces periodicity of the resolution of simple modules to almost periodicity for the algebra.

(7) Let the resolution of A_{mon} be locally infinite. Then the resolution of every locally equivalent algebra B_{mon} is infinite. This includes all $B_{mon} \in [A]_{mon}$. The same for the resolutions of the simple B_{mon}-modules and hence of the simple B-modules. With Happel, the resolution of A is locally infinite. $\qquad\square$

Index

Bibliography

[1] D. J. Anick. On the homology of associative algebras. *Trans. AMS*, 296:641–659, 1986.

[2] D. J. Anick and E. L. Green. On the homology of quotients of path algebras. *Communications in Algebra*, 15(1& 2):309–341, 1987.

[3] I. Assem, D. Simson, and A. Skowroński. *Elements of the representation theory of associative algebras*, volume 1. London Mathematical Society, Student Texts 65, 2006.

[4] M. J. Bardzell. The alternating syzygy behavior of monomial algebras. *Journal of Algebra*, 188:69–89, 1997.

[5] H. Cartan and S. Eilenberg. *Homological algebra*. Princeton University Press, Princeton, NJ, 1965.

[6] K. Erdmann and A. Skowroński. Periodic algebras. In *Trends in representation theory of algebras and related topics*, pages 201–251, 2008.

[7] D. R. Farkas, C. D. Feustel, and E. L. Green. Synergy in the theory of gröbner bases and path algebras. *Can. J. Math.*, 45(4):727–739, 1993.

[8] E. L. Green, D. Happel, and D. Zacharia. Projective resolutions over artin algebras. *Illinois J. Math.*, 29(1):180–190, 1985.

[9] D. Happel. Hochschild cohomology of finite-dimensional algebras. *Lecture Notes in Math. 1404, Springer, Berlin*, pages 108–126, 1989.

[10] J. A. de la Peña and C. Xi. Hochschild cohomology of algebras with homological ideals. *Tsukuba J. Math.*, 30(1):61–79, 2006.

[11] S. König and H. Nagase. Hochschild cohomology and stratifying ideals. *J. Pure Appl. Algebra*, 213(5):886–891, 2009.

[12] R. S. Pierce. *Associative algebras*. Springer Verlag, 1982.

[13] M. Suárez-Alvarez. Applications of the change-of-rings spectral sequence to the computation of Hochschild cohomology. *arXiv:0707.3210*, 2007.